Geology: A Synopsis
Part I: Physical Geology

J. Allan Cain
Eugene J. Tynan

University of Rhode Island

KENDALL/HUNT PUBLISHING COMPANY
Dubuque, Iowa, USA • Toronto, Ontario, Canada

Printed in the United States of America

B   402162   01

Contents

Preface

This book evolved from our experiences teaching introductory courses in Physical Geology, Historical Geology, and a combined version called Geological Earth Science.

Our intent is to provide a nucleus of information which the instructor can supplement through lecture material and outside readings. In this respect, the book represents a reversal in the roles mandated by several of the newer texts which require the instructor to simplify, summarize and select from the wealth of material presented. Using this book, the instructor will provide additional insights, examples, and readings to supplement the abbreviated text.

For the student, the book should provide a good starting set of notes emphasizing fundamental concepts and interrelationships in geology, as well as basic information and vocabulary. To increase its usefulness, the book also contains a Glossary defining some 600 words likely to be encountered in lectures and readings.

Some material is best treated in laboratory sessions and/or presented through the use of color slides and examination of specimens in class. We, therefore, have not included many photographs of minerals and rocks, or examples of topographic and geologic maps; several excellent laboratory manuals, as well as sets of maps and air photographs, are readily available.

There are many sources of information on topics we mention only briefly or do not cover at all; some are listed on page x as Additional Readings.

In our combined 40 years of teaching we have, of course, absorbed information and approaches that reflect interactions with teachers, colleagues and students. We also have been influenced by published material, including texts used over the years. In preparing this book we have been greatly assisted by our colleagues: Part I has been reviewed by Jon C. Boothroyd, Geoffrey C. Cain, John J. Fisher, Reinhard K. Frohlich, O. Don Hermes (who read several versions), Charles F. Hickox, Sally A. Holmes, William F. MacLean (who also provided copies of his own course summaries), and Harriet E. Powell. In addition, Roger E. Dulude, Jr. is responsible for the photography; Harriet Powell prepared several of the illustrations and provided invaluable general assistance. Susan J. Morrow and Phyllis D'Abate cheerfully typed the many versions of the manuscript.

We are indebted to them all; inadequacies in our treatment, however, are solely our responsibility.

GEOLOGY: A SYNOPSIS, PART II will place special emphasis on *Historical Geology*. The authors anticipate a mid-1980 publication date.

Rish my vec as ayns cooinaghtyn my voir as ayr lesh graih.

J.A.C.

To the Reader

Through the study of geology we hope you will become aware of:

a. the excitement and challenge of seeking to understand the earth, its processes, evolution, and limits.

b. the tremendous range of scale in the natural world, the cyclicity of natural processes, the impermanence of physical things, the constancy of change, and the enormity of time.

c. the factors of the physical environment which influence man's structures and well-being, and the scientific principles which govern them. These factors include hazards such as floods, landslides, shore erosion and deposition, earthquakes, etc.

d. the ways in which man affects the environment by modifying the natural regime and changing the rates of natural processes.

e. the uniqueness of the formation of natural resources and the resulting limitations on society's use, and abuse, of them.

f. the origins of materials and landscapes, as well as ways of deciphering the earth's physical and biological evolution.

g. the questions to ask various special-interest groups in order to assess the possible impact of their proposals on the earth and society.

Additional Readings

These are only some of the many good sources of further information that are available.

General Geology

Foster, R. J. 1978. *General Geology* (Third Edition). Columbus: Charles E. Merrill.
Press, F., and R. Siever (compilers). 1975. *Planet Earth: Readings from Scientific American*. San Francisco: W. H. Freeman.
Stokes, W. L., S. Judson, and M. D. Picard. 1978. *Introduction to Geology* (Second Edition). Englewood Cliffs: Prentice-Hall.

Physical Geology

Davis, S. N., P. H. Rietan, and R. Pestrong, 1976. *Geology—Our Physical Environment.* New York: McGraw-Hill.
Flint, R. F., and B. J. Skinner, 1977. *Physical Geology* (Second Edition). New York: John Wiley and Sons.
Hamblin, W. K. 1978. *The Earth's Dynamic Systems* (Second Edition). Minneapolis: Burgess.
Holmes, Arthur. 1978. *Holmes Principles of Physical Geology* (Third Edition, revised by D. L. Holmes). Halsted Press, Division of John Wiley and Sons.
Judson, S., K. S. Deffeyes, and R. B. Hargraves. 1976. *Physical Geology.* Englewood Cliffs: Prentice-Hall.
Leet, L. D., S. Judson and M. E. Kauffman. 1978. *Physical Geology* (Fifth Edition). Englewood Cliffs: Prentice-Hall.
Press, F., and R. Siever. 1978. *Earth* (Second Edition). San Francisco: W. H. Freeman.
Sawkins, F. J., C. G. Chase, D. G. Darby and George Rapp, Jr. 1978. *The Evolving Earth* (Second Edition). New York: Macmillan.
Shelton, J. S. 1966. *Geology Illustrated.* San Francisco: W. H. Freeman.

Historical Geology (See also, Part II)

Dott, R. H., Jr., and R. L. Batten. 1976. *Evolution of the Earth* (Second Edition). New York: McGraw-Hill.
Eicher, D. L. 1976. *Geologic Time* (Second Edition). Englewood Cliffs: Prentice-Hall.
Seyfert, C. K., and L. A. Sirkin. 1979. *Earth History and Plate Tectonics* (Second Edition). New York: Harper and Row.
Stearn, C. W., R. L. Carroll, and T. H. Clark. 1979. *Geological Evolution of North America* (Third Edition). New York: John Wiley and Sons.

Plate Tectonics and the Earth's internal processes

Bolt, B. A. 1978. *Earthquakes—A Primer*. San Francisco: W. H. Freeman.

Hallam, A. 1973. *A Revolution in the Earth Sciences*. Oxford: Clarendon Press.

LePichon, X., J. Francheteau, and J. Bonnin. 1973. *Plate Tectonics*. New York: Elsevier. (Reference-level treatment.)

Meyerhoff, A. A., and H. A. Meyerhoff. 1972. "The New Global Tectonics: Major Inconsistencies," American Association of Petroleum Geologists Bulletin, v. 56, pp. 269-336.

National Association of Geology Teachers. 1979. Crustal Evolution Education Project (CEEP). Rochester: Ward's Natural Science Establishment. (Over 30 modular, pre-college units on sea-floor spreading, continental drift and plate tectonics.)

Uyeda, Seiya. 1978. *The New View of the Earth*. San Francisco: W. H. Freeman.

Wilson, J. T. (compiler). 1970. *Continents Adrift—Readings from Scientific American*. San Francisco: W. H. Freeman.

Wilson, J. T. (compiler). 1976. *Continents Adrift and Continents Aground: Readings from Scientific American*. San Francisco: W. H. Freeman.

Wyllie, P. J. 1971. *The Dynamic Earth: Textbook in Geosciences*. New York: John Wiley and Sons. (Upper-level text.)

Wyllie, P. J. 1976. *The Way the Earth Works*. New York: John Wiley and Sons.

Minerals and rocks

Ernst, W. G. 1969. *Earth Materials*. Englewood Cliffs: Prentice-Hall.

Laporte, L. F. 1979. *Ancient Environments* (Second Edition). Englewood Cliffs: Prentice-Hall.

Macdonald, G. A. 1972. *Volcanoes*. Englewood Cliffs: Prentice-Hall.

Prinz, M., G. Harlow, and J. Peters (editors). 1977. *Simon and Schuster's Guide to Rocks and Minerals*. (Color photographs of minerals and rocks.)

Mineral and energy resources

Cook, E. 1976. *Man, Energy, Society*. San Francisco: W. H. Freeman.

Park, C. F., Jr. 1975. *Earthbound: Minerals, Energy, and Man's Future*. San Francisco: Freeman, Cooper and Co.

Ruedisili, L. C., and M. W. Firebaugh (editors). 1978. *Perspectives on Energy* (Second Edition). New York: Oxford University Press.

Skinner, B. J. 1976. *Earth Resources* (Second Edition). Englewood Cliffs: Prentice-Hall.

Environmental Geology

Bolt, B. A., W. L. Horn, G. A. Macdonald, and R. F. Scott. 1975. *Geological Hazards*. New York: Springer-Verlag.

Cargo, D. N., and B. F. Mallory. 1977. *Man and His Geologic Environment* (Second Edition). Reading, Mass.: Addison-Wesley.

Griggs, G. B., and J. A. Gilchrist. 1977. *The Earth and Land Use Planning.* North Scituate, Mass.: Duxbury Press.

Keller, E. A. 1979. *Environmental Geology* (Second Edition). Columbus, Ohio: Charles E. Merrill.

Oceanography

Bhatt, J. J. 1978. *Oceanography—Exploring the Planet Ocean.* New York: D. Van Nostrand Co.

Drake, C. L., J. Imbrie, J. A. Knauss, and K. K. Turekian. 1978. *Oceanography.* New York: Holt, Rinehart and Winston.

Gross, M. G. 1977. *Oceanography—A View of the Earth.* (Second Edition). Englewood Cliffs: Prentice-Hall.

Planetary Geology

Short, N. M., 1975. *Planetary Geology.* Englewood Cliffs: Prentice-Hall.

1. Overview: Materials, Appearance, Cycles

Although in many cases its importance is unappreciated, geology affects our lives in numerous ways. For example:

Hazards. We need to understand, predict and, if possible, control natural disasters such as earthquakes, landslides, floods, volcanic eruptions. Engineering projects (dams, sanitary landfills, highways, coastal structures, buildings, etc.) should be sited in geologically sound locations. We also should understand man's interactions with the natural regime, including climatic modifications, pollution, and related effects. Similarly, we should attempt to achieve optimal use of the land by an ever-growing population. The study of these aspects is called *Environmental Geology* (and the more specialized *Engineering Geology*).

Resources. Industrialized society exists because of available mineral resources. These include **energy sources**—oil, gas, coal, uranium, geothermal (steam and hot water), oil shale; **metals**—iron and the steel-making ferroalloy elements, base metals such as aluminum and copper, as well as precious metals, and minor metals; **non-metals**—building materials, industrial minerals, fertilizer minerals, gemstones, and water. This branch of geology, and its complex interrelationships with economics, politics, technology, and others, is called *Economic Geology.*

In addition to these two very practical reasons for knowing something about geology there are other considerations:

Aesthetics. In understanding the origin of gemstones, landscapes such as the Grand Canyon (Fig. 1.1), and other natural features, one can gain a greater appreciation of nature and man's relationship to it. Similarly, through an awareness of the enormity of geologic time and the vast physical and biological changes that have occurred on earth during the past 4.7 billion years, we gain a perspective on our own existence.

In the words of Thomas H. Huxley: "To a person uninstructed in natural history, his country or seaside stroll is a walk through a gallery filled with wonderful works of art, ninetenths of which have their faces turned to the wall."

Scientific. To those interested in explaining natural phenomena through an understanding of fundamental laws, principles and hypotheses, geology offers a significant challenge. Although in many cases we can observe and measure present geologic processes (rivers, volcanoes, glaciers, etc.) and we can simulate natural events through controlled laboratory experiments (for example, high-temperature and pressure equipment to determine the behavior of materials hundreds of kilometers within the earth), in much of geology we are presented only with the fragmentary results of natural events that occurred millions of years

1

Figure 1.1. The Grand Canyon, Arizona. The rocks at the bottom of the canyon, carved by the Colorado River, are some 2 billion years old. The horizontal layers record 350 million years of the earth's history. (Photo courtesy U. S. Department of the Interior, National Park Service).

ago. Geologists then must seek to discern patterns and relationships in the data and develop models for the earth's behavior that would produce these observed results and still be consistent with fundamental scientific principles. Similar approaches are used in the study of planetary geology. Plate tectonics (Fig. 1.10) is an example of how geologic, physical, chemical, and biological concepts and information have been synthesized into a revolutionary hypothesis that explains the distribution of mountain ranges, rock-types, mineral resources, earthquakes, volcanoes, biological species, etc.

The Geologic Record

To understand the evolution of the earth it is necessary to visualize processes acting slowly over millions or even billions of years. The physical record consists of:

1. *Morphology* (or appearance) of the earth. This can vary from a global scale, such as continents and ocean basins, to an extremely local scale where we consider only a single beach, valley or layer of rock. Features of all scales and types consist of regolith and/or bedrock. Their particular shapes (such as glacial moraines, volcanic cones, river valleys) are the result of certain processes operating within and on the earth (page 8).

2. *Regolith.* This is loose, unconsolidated material familiar to all as beach sand, soil and the like. Regolith can be compacted and cemented into bedrock; conversely, bedrock can be weathered and broken down to regolith.

3. *Bedrock.* Apart from a thin, discontinuous layer of regolith, the earth consists of three types of bedrock.

Igneous rock—originating from molten material called magma (as from volcanoes, producing basalt, or cooled underground to give granite).

Sedimentary rock—derived from fragments of pre-existing rocks (for example, compacted and cemented sand giving sandstone), precipitated from solution (such as limestone or rock salt), or accumulated from organic material (like coal).

Metamorphic rock—an existing rock that has been modified (or "changed form"), usually by increased temperature and pressure (for example: limestone recrystallized and changed to marble).

All three rock types and regolith consist of minerals (Table 2.2) and are related through the rock cycle (Fig. 1.12).

Major Surface Features of the Earth

Major features (morphology) of the earth (Fig. 1.2) are continents (29% of the earth's surface), ocean basins (60%), continental shelves and slopes (11%).

a. *Continents* (which are concentrated in the northern hemisphere) consist of shields, platforms, and belts of folded mountains (Fig. 1.3):

Shields are flat, geologically old regions of Precambrian (Table 4.4) igneous and metamorphic rocks (older than 600 million years).

Platforms are veneers of nearly horizontal layers of younger sedimentary rocks which overlie shield-type material.

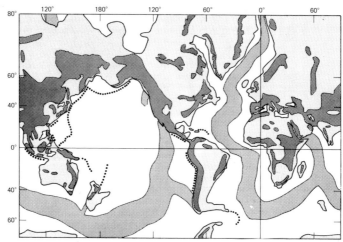

Figure 1.2. Major surface features of the earth. Key: white, abyssal ocean floor; medium shading, oceanic ridge system; heavy dotted lines, oceanic trenches; light shading, continental platform and continental shelf; dark grey, mountains, intermontane basins, associated hills, and some elevated plateaus. (From *The Dynamic Earth*, Peter J. Wyllie. John Wiley & Sons, Inc., New York, Copyright, 1971.)

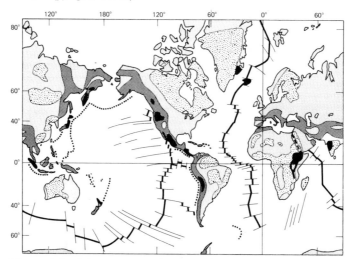

Figure 1.3. Major tectonic features of the earth. Key: heavy lines, active rift systems of oceanic ridges; light lines, oceanic faults; dotted lines, oceanic trenches; light shading, continental platforms; ornamented, continental shields; dark grey, Tertiary (2-65 million years old) fold mountain chains; black, Cenozoic (up to 65 million years old) volcanic regions. (From *The Dynamic Earth*, Peter J. Wyllie. John Wiley & Sons, Inc., New York, Copyright, 1971.)

Platforms and shields together form the stable interior of continents, called the **craton**.

Linear belts of **folded mountains,** which are produced by compression during collision of lithospheric (p. 7) plates, generally extend along the margins of continents. Continents appear to have grown through time by the addition of these belts at their margins. Folded mountain ranges consist of all three rock types. One major linear belt which extends north-south includes the Rockies and Andes. Another includes the Alps and Himalayas and is generally east-west. Sedimentary layers in these belts typically show deformation by folding and faulting (Section 16). Major mountain ranges that exist today are geologically young, whereas the shields represent very old ranges that have been worn flat by erosional agents. Folded ranges are the result of orogeny (Section 16).

In general, the continents consist of material, mostly granitic (Table 3.2) in composition, that is 30 to 60 km thick and less dense than the basaltic material (5-10 km thick) which is typical of the ocean basins. Rocks 3.8 billion years old have been identified on the continents (e.g., west Greenland and Minnesota); the rocks in the ocean basins are less than 200 million years old.

b. *Ocean basins* are separated from the continents by the **continental shelves**, which are simply the submerged margins of the continents (Fig. 1.4). As sea level fluctuated throughout geologic time (a recent cause is glacial advance and retreat) the size of the shelves also varied.

Figure 1.4. Features of the Atlantic Ocean basin (Courtesy, ALCOA).

Connecting the shelves to the deep ocean basins are the **continental slopes and rises** (Fig. 14.3), which are cut, in some areas, by **submarine canyons** (Fig. 14.1).

Deep ocean features have a generally basaltic composition and are blanketed by a thin layer of clay and organic ooze. Specific oceanic topographic and tectonic units (Fig. 1.3) include:

Ridges—a world-encircling chain (60,000 km long) of submarine mountains with central tensional rifts, believed to be spreading centers where new basaltic material is added to the lithospheric plates (page 13) as part of the plate tectonic cycle (Fig. 1.10). Ridge segments are offset by transform faults (Figs. 1.4 and 16.18).

Trenches—the deepest parts of the ocean floor (as much as 11 km below sea level), found adjacent to volcanic island arcs or continental mountain ranges. Here, old lithosphere is consumed as it is carried downward as part of the sinking ocean plate, at the opposite end of a conveyor-belt-like relationship to the lithosphere-producing ridges. Some 100-400 km from the trenches (usually toward the nearby continent) major volcanic and deep-focus earthquake activity occur (Fig. 1.11). (*Marginal basins*, or seas, with thick accumulations of sediment, form between island arcs and continents; e.g., Japan Sea.)

Other features of ocean basins include the abyssal floors, made up of **abyssal hills** and **abyssal plains**. The plains are generally flat areas of basalt hills completely covered by sediment. Also found in the ocean basins are thousands of **seamounts** (usually extinct volcanic cones, some of which are in linear belts), and **guyots** which are flat-topped volcanic cones produced by wave erosion when sea level was lower, and/or the cones stood higher (Fig. 14.2) before being moved away from the ridge by sea-floor spreading (page 13).

It might be noted that the Pacific Ocean itself accounts for over 33% of the earth's surface area (cf. less than 30% for all the continents).

Interior of the Earth

We now have some idea of the earth's major surface features. To complete our overview let us look at the interior of the earth, illustrated in Fig. 1.5.

Core—Predominantly iron (and nickel), forming a solid inner core and liquid outer core (evidence for its liquid nature is the fact that shear waves from earthquakes do not pass through it). The liquid outer core (coupled with the earth's rotation) is thought to be responsible for the earth's magnetic field. The temperature at the core-mantle boundary probably is around 3700°C. Density of the outer core is approximately 9 grams per cubic centimeter (g/cc); at the earth's center it is approximately 12 g/cc. The density of the earth as a whole is over 5.5 g/cc.

Mantle—Density 4.5 g/cc, ranging from 3 at the top to 6 at the bottom; magnesium silicates (minerals, olivine, pyroxene, garnet, etc.; rock, peridotite) in the outer part, iron-magnesium silicates and oxides in the deeper part. The mantle accounts for over 80% of the volume of the earth and consists of several concentric zones (marked by changes in the velocity of earthquake [seismic] waves). Among the most important of these zones is the **asthenosphere**, some 600 km thick, lying generally less than 100 km below the earth's surface. Temperatures in this zone range from approximately 1000°C at the top to 2000°C at the bottom (at a depth of some 700 km). The upper 150 km of the asthenosphere is a zone of low seismic wave velocity, called the *low velocity layer*, which behaves plastically, permitting movement of the overlying lithosphere plates (*plate tectonics*). Above the asthenosphere is the rigid **lithosphere** (70-150 km thick), the lower part of which is the outermost part of the mantle.

Crust—The upper lithosphere (Fig. 1.5) consists of the **oceanic** (density 2.9 g/cc) and **continental** (density 2.8 g/cc) **crust**, (sometimes called *sima* and *sial*, respectively) separated from the mantle by the **Mohorovicic discontinuity**, or Moho—a zone below which the fastest type of earthquake waves (P waves) increase in velocity from 6-7 km/sec to 8 km/sec. This discontinuity probably reflects the difference in chemical composition between mantle peridotite and crustal basalt and granite. (An alter-

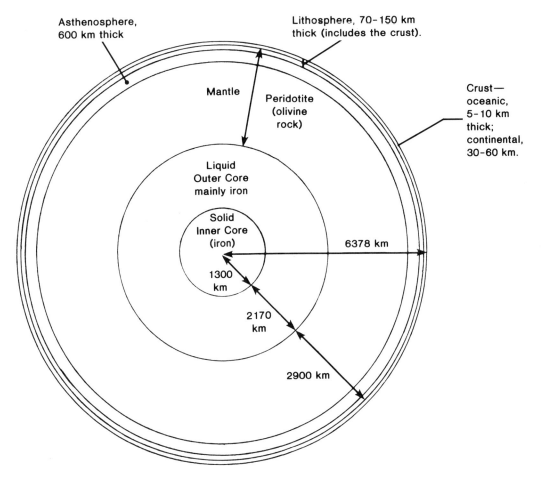

Figure 1.5. Cross-section showing the major concentric shells of the earth's interior.

native explanation is that the chemical composition of the upper mantle is the same as that of basalt but, because of the high temperature and pressure, different minerals are developed, forming the rock eclogite; these mineral [phase] changes could affect the velocity of seismic waves.)

Cycles and Energy Sources

To complete our overview, it is necessary to consider the cycles and energy sources operating on and within the earth.

The earth consists of **matter** or groups of atoms. Three states of matter are recognized: gas (*atmosphere*), liquid (*hydrosphere*), and solid (*lithosphere*). The earth's energy-transfers change matter from one state to another (liquid magma to solid igneous rock, and vice-versa).

The earth is a dynamic planet. Its morphology changes through time and represents the constant battle between **constructive or internal processes**, where new mountains and other topographic features are produced (volcanoes [Fig. 1.6], folds, faults), and **destructive or external processes** which attempt to level the earth through erosional activity (Fig. 1.7) by

Figure 1.6. Surtsey, south of Iceland. An island created by volcanic activity in 1963. (Photo, Gerald J. Daub.)

Figure 1.7. Bryce Canyon, Utah, showing horizontal sandstone, limestone and shale layers eroded by water and wind along joints and bedding planes. (Photo courtesy of U. S. Department of the Interior, National Park Service.)

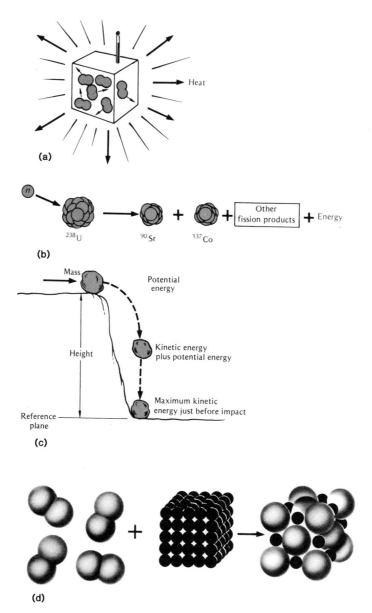

Figure 1.8. Various forms of energy. (A) Thermal energy caused by agitation or vibration on an atomic or molecular scale, (B) Nuclear energy caused by certain associations of subatomic particles, (C) Potential energy converted to kinetic energy by gravitational attraction of the earth on the mass, (D) Atoms of chlorine gas as molecules, Cl_2, plus atoms of sodium metal, Na, combine to form salt, NaCl, and release heat. The reaction causes chemical energy to be changed to thermal energy. (From *Geology: Our Physical Environment*, by S. N. Davis, P. H. Reitan and Raymond Pestrong. McGraw-Hill, Inc., New York, Copyright 1976).

wind, rivers, ice, waves, etc. (If only the latter processes had operated, the earth would be a smooth, water-covered planet.)

Energy sources for these processes, which are responsible for producing the different types of bedrock and regolith as well as the earth's morphology, involve the earth's internal thermal energy, including *radioactive decay* of certain elements (internal processes), and *solar, gravitational* and *tidal* energy (external processes).

The influence of these energy sources and the transfers among energy types (thermal, kinetic, potential, as illustrated in Fig. 1.8) can be seen in **three major geologic cycles:**

Hydrologic Cycle

Figure 1.9 illustrates the hydrologic cycle. Note that the oceans contain over 97% of the water in the hydrosphere. Glaciers contain over 2%, thus the water available to us in rivers, lakes and underground aquifers is less than 0.7% of the total hydrosphere.

Note also that the continents have a net gain at the expense of the oceans as they receive 25% of the water precipitated but provide just 15% of the water evaporated. This precipitation eventually returns to the oceans primarily as rivers and groundwater. Movement of wa-

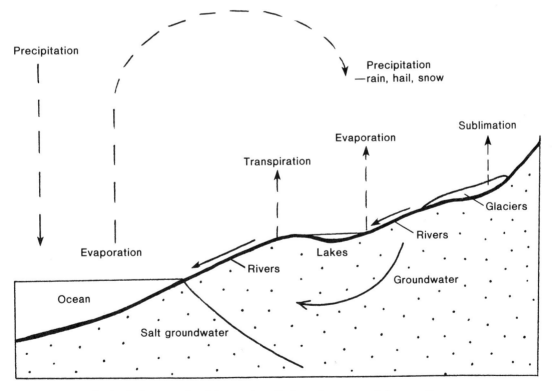

Figure 1.9. The hydrologic cycle, driven primarily by solar energy (plus gravity and the earth's rotation).

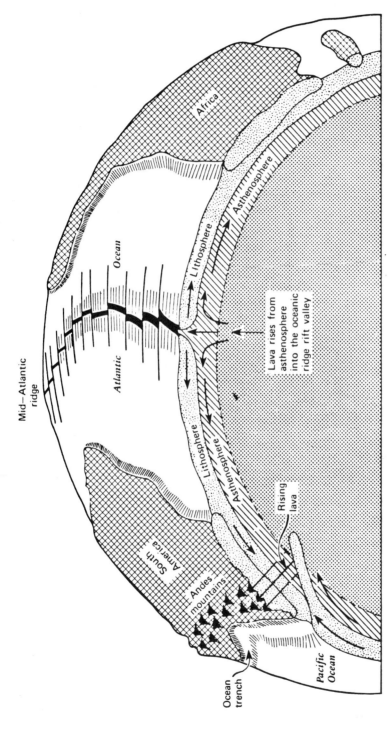

Figure 1.10. The plate tectonic cycle: lithosphere is added at spreading ridges and consumed in trenches (subduction zones). (From *The Way the Earth Works* by Peter J. Wyllie. John Wiley and Sons, Inc., New York, Copyright 1976.)

ter in the hydrologic cycle is an essential part of the rock cycle, as it transports fragments and dissolved material, and cements regolith into sedimentary bedrock.

Tectonic Cycle

The tectonic cycle (Fig. 1.10) explains the distribution of tensional and compressional features in the earth, volcanic chains, earthquakes, folded mountain belts, etc. (See also, Section 18.) The tectonic cycle also provides a mechanism for carrying surface materials down into the mantle (at converging plate boundaries or trenches) and bringing mantle material to the surface (at diverging plate boundaries or ocean ridges). Thus it is an essential part of the rock cycle (as shown in Fig. 1.14). Figure 1.10 is a diagrammatic cross-section of the tectonic cycle; the distribution of these features around the earth is shown in Figs. 1.3 and 1.11.

The earth's surface consists of 7 major and several smaller rigid plates of lithosphere (Fig. 1.11). When these plates interact they produce **divergent ridges** (sea-floor spreading) at one margin, **convergent trenches** (subduction zones) at the opposite margin, and sideways-slipping **transform faults** which offset the ridges and permit one plate to slide past another. (See also Figs. 3.17, 3.18, 3.19, 3.21, 5.1, 6.2, 16.22.)

It should be clear that plate boundaries are extremely important in the study of geologic hazards because earthquakes and volcanoes are located mainly along the margins of plates.

Table 1.1 is an outline of the key characteristics of plate boundaries which will be referred to repeatedly throughout this book. Note that there are different sorts of geologic processes and products depending, for example, on whether collision occurs between two plates with no continents involved, or between two continent-bearing plates. The former would give trenches and volcanic **island arcs** like Japan and the Philippines; the latter, non-volcanic mountain ranges like the Himalayas and Alps.

Rock Cycle

The rock cycle, shown in Fig. 1.12, results from the operation of the hydrologic and tectonic cycles. External (destructive) processes involve primarily the hydrologic cycle (driven by solar energy, gravity, earth's rotation); internal (constructive) processes are the result of the tectonic cycle, driven by the earth's internal heat (radioactive decay, etc.). These relationships are shown in a more comprehensive representation of the rock cycle in Fig. 1.13. Fig. 1.14 is a generalized illustration of how particular types of igneous, sedimentary, and metamorphic rocks are related in space and time through the tectonic cycle.

These and other cycles, such as the climatic or various geochemical cycles like that of carbon, appear to have operated throughout the earth's history. Fundamental to the study of geology is the principle that the earth's processes acting today have been acting throughout geological time. In other words, by studying present geologic processes and products we can **understand** past geologic events and the origin of the geologic record. This is the **Principle of Uniformitarianism.** Changes in the physical and biological records through time and the unraveling of earth history form the material in the second part of this book.

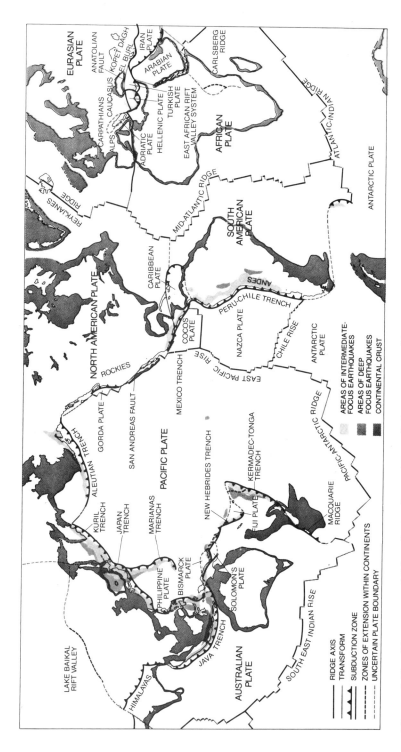

Figure 1.11. The system of lithospheric plates. (From "*Plate Tectonics*" by J. F. Dewey, Copyright 1972, by Scientific American, Inc. All rights reserved.)

TABLE 1.1   Classification of Plate Boundaries

Boundary	Ocean-ocean	Ocean-continent	Continent-continent
Divergent (tension)	Ridge crest S earthquakes,* narrow belt. Submarine lavas		Rift valley S earthquakes, wide zone. Volcanoes
Convergent (compression)	Ocean trench and volcanic island arc SID earthquakes, wide belt Volcanoes	Ocean trench and young mountain range SI±D earthquakes, wide belt Volcanoes	Young mountain range S ± I earthquakes, wide zone No volcanoes
Transform (neither tension nor compression)	Fracture zone of ridges and valleys. S earthquakes, narrow belt only between offset ridges No volcanoes		Fault zone S earthquakes, broad zone No volcanoes

*S, shallow focus, less than 70 km deep
I, intermediate focus, 70-300 km
D, deep focus, greater than 300 km

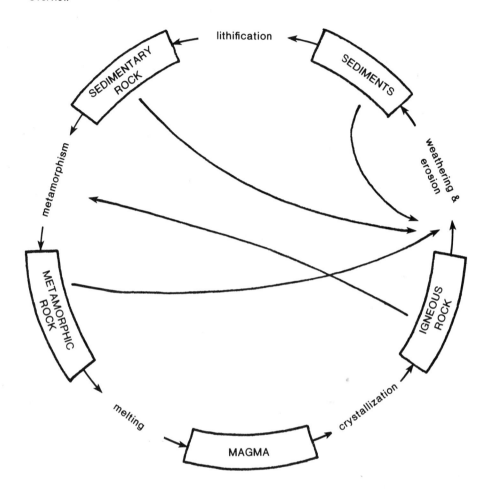

Figure 1.12. Simplified diagram of the rock cycle. A more complete version is given in Figure 1.13.

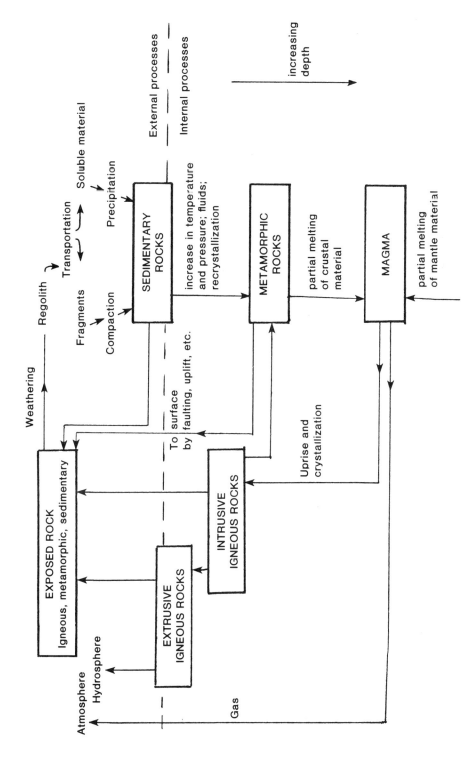

Figure 1.13. The rock cycle. A simplified version is given in Figure 1.12. See also, Figure 1.14.

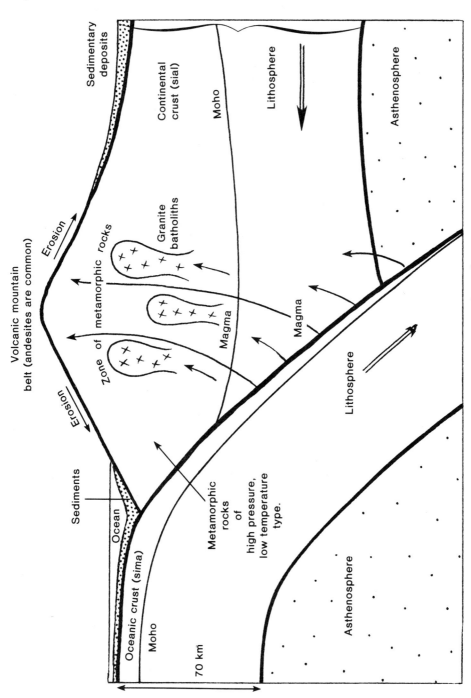

Figure 1.14. Simplified cross-section showing a relationship between the rock cycle and tectonic cycle. Sediments in the trench are subducted, metamorphosed, melted; magma rises into the crust causing metamorphism and volcanic activity. Weathering and erosion of the volcanic material gives new sediment. See also, Figure 1.13.

MATERIALS OF THE EARTH

Crystals of galena (top) and calcite. Ward's Natural Science Establishment, Inc.

Materials of the Earth

This outline is followed by a more detailed discussion of minerals and rocks.

1. *Minerals*—are naturally occurring, inorganic solids with atoms bonded into a definite internal structure. They have diagnostic physical properties, and chemical compositions that vary only within specific limits. 95% of the earth's crust consists of silicate minerals, with the rest being oxides, carbonates, sulfates, sulfides, etc.

2. *Rocks*—are rigid aggregates of minerals (Fig. 2.1A & B). The size, shape, arrangement (texture), and composition of minerals permit interpretation of the origin and history of the rock.

    a. *Igneous rocks* (65% of the volume of the crust) form from the cooling of magma (molten rock material) which originates by partial melting of the lower continental crust (to give mainly granitic magmas), or of the upper mantle (mainly basaltic magmas).

    There is a generally predictable sequence in which minerals crystallize from magma as it cools (Fig. 3.15).

    Extrusive igneous rocks, consisting of small crystals or glass, result from rapid cooling of magma (lava) when it flows on the surface (e.g., basalt). Intrusive rocks cool slowly beneath the earth's surface thus developing larger crystals (e.g., granite).

    Basaltic magmas (low in silica and high in iron) have low viscosity (resistance to flow) which produces non-explosive shield volcanoes and fissure flows, as well as intrusions in the form of dikes (discordant) and sills (concordant). Granitic magmas (high in silica and low in iron) generally have high viscosity. This high viscosity, together with a higher content of volatiles, produces explosive volcanoes (composite cones, rhyolitic welded tuffs). Most economically important metals are associated with granite batholiths (the largest intrusive igneous masses, having an exposed surface area of at least 100 square kilometers).

    The distribution of igneous rocks is closely related to plate tectonic boundaries (see Fig. 3.6).

    b. *Sedimentary rocks* (less than 10% of the volume of the crust) are derived from material weathered from pre-existing rocks; the type and extent of weathering depend to a great degree on rainfall (and temperature) thus the hydrologic cycle is extremely important. This relationship is reflected in the fact that most sediments are deposited ultimately in the oceans (transported there by rivers, underground water, glaciers, etc.).

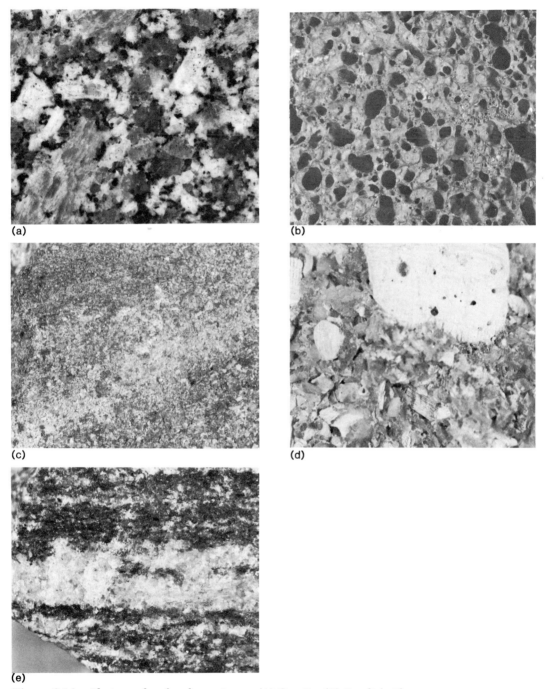

(a)

(b)

(c)

(d)

(e)

Figure 2.1A. Photographs of rock specimens. (A) Granite, (B) Basalt (with gas holes or vesicles), (C) Sandstone, (D) Coquina (limestone made of shell fragments), (E) Gneiss.

(a)

(b)

(c)

Figure 2.1B. Photographs of thin-sections of rocks. (A) Granite—interlocking quartz, feldspar and biotite, (B) Sandstone—rounded to angular grains of quartz and feldspar in a fine-grained matrix, (C) Gneiss—aligned biotite, together with quartz and feldspar. (Photo by O. Don Hermes)

One class of sedimentary rocks, the clastic or detrital, consists of transported mineral or rock fragments; the other class, chemical, consists of chemical precipitates and/or organic material.

The type of sedimentary rock produced is determined by the material at the source area, the nature and extent of weathering, the type of transporting (erosional) agent, and environmental conditions at the site of deposition.

c. *Metamorphic rocks* (less than 30% of the volume of the crust) result when increased temperature and pressure, perhaps aided by circulating fluids, change pre-existing rocks and minerals without major change in the bulk chemical composition of the rock unit. (Where bulk compositional changes occur the process is called metasomatism.)

Reactions occur in the solid-state and may involve recrystallization of existing minerals and/or growth of new minerals, typically with a preferred orientation in the direction of least stress (Fig. 2.1B). The common result is a planar arrangement of minerals, called foliation (e.g., in slate, schist and gneiss).

Regionally-metamorphosed rocks commonly are foliated; contact metamorphism produces a non-foliated rock texture called hornfels (directional stress is not significant).

Regional metamorphism affects large areas and is associated with plate boundaries, especially in folded mountain belts during plate collisions. Contact metamorphism involves local modifications around igneous intrusions.

3. *Regolith*—produced by mechanical and chemical weathering of pre-existing rocks. Climate ultimately affects the type of regolith more than the original rock-type does. Residual regolith develops a soil profile; transported regolith becomes sedimentary rock as part of the rock cycle.

2. Minerals

Significance of Minerals

Apart from their economic significance as sources of essential metals and raw materials for construction, agriculture, industry and other purposes (see page 81), minerals are keys to interpreting the earth.

Minerals give information, for example, about the composition of the parent magma and where it originated; the temperatures and pressures of metamorphism and the nature of the material before it was metamorphosed; the type of source area that was eroded to give sedimentary rocks; if weathering was rapid or slow; the erosional agent that transported a particular sediment; the salinity, acidity (pH), and oxidizing-reducing conditions (Eh) in which material precipitated.

To make these and similar interpretations, geologists may study the behavior of minerals or synthetic materials under simulated conditions in the laboratory. It is possible, for example, to reproduce the conditions existing more than 700 km within the earth. Similarly, using the dictum of uniformitarianism (that the present is the key to the past) modern processes and their products (volcanoes, beaches, coral reefs, etc.) can be used as analogs for ancient rocks and their mineral assemblages. Further, by studying the abundance of certain parent/daughter isotopes (e.g., of uranium and lead) it is possible to establish the age of a mineral and, hence, the rock age. Magnetization of certain iron-rich minerals is retained in old rocks and this permits determination of ancient magnetic polarities and pole positions (of great significance in sea-floor spreading and plate tectonics). Also, minerals preserved in the form of plant and animal fossils have their own significance in establishing ancient ecological preferences, climatic changes, evolutionary patterns, etc., as well as permitting correlation of widely separated rock assemblages and the development of a relative time scale (Table 4.4).

Origin and Types of Minerals

Minerals are the common solid form of matter. There are some 2200 different minerals but only 30 or so common rock-forming ones. Minerals form when atoms combine by bonding in an organized lattice arrangement (e.g., Fig. 1.8) a) **from a liquid,** such as crystallization of a cooling magma (*igneous processes*) or precipitation from aqueous solutions (*chemical sedimentary processes*); b) **from a pre-existing solid** by alteration of existing minerals at normal temperatures and pressures (*weathering*) or at elevated temperatures and pressures (*metamorphism*); c) **from a gas** by condensation of vapor (sulfur around volcanic vents).

Silicon combines readily with oxygen (Table 2.1) forming silicate minerals which make up some 95% of the earth's crust—from the familiar and chemically simple quartz (SiO_2), to quite complex combinations of other elements in addition to silicon and oxygen, e.g., amphibole, $Ca_2Mg_5 Si_8O_{22} (OH)_2$ (tremolite).

Mineral groups other than silicates include oxides and hydroxides, carbonates, sulfates, sulfides, phosphates, halides, and some elements that occur in the native state (uncombined with any other element). A summary of the major groups of minerals and their properties is given in Table 2.2. Table 2.3 indicates the volumetrically important minerals in the earth's crust.

TABLE 2.1   Major Elements in the Earth's Crust (weight %)

Oxygen (O)	46.6
Silicon (Si)	27.7
Aluminum (Al)	8.1
Iron (Fe)	5.0
Calcium (Ca)	3.6
Sodium (Na)	2.8
Potassium (K)	2.6
Magnesium (Mg)	2.1
	98.5

The abundance of elements given in Table 2.1, shows that almost all the metals on which we depend for our industrialized society (copper, lead, zinc, gold, uranium, nickel, etc., etc.) together make up only 1% of the earth's crust.

The more abundant metallic elements in the crust (iron, aluminum, titanium, manganese, and magnesium) typically occur as oxides, hydroxides, and/or carbonates. The scarcer elements (less than 0.1%) commonly substitute for other elements in various minerals. When they form their own minerals they are mostly sulfides. Some, however, such as gold and platinum, occur as native elements; a few others (chromium, tin, tungsten, uranium, vanadium) form oxygen-containing minerals.

There is virtually an infinite number of *proportions* in which minerals can be combined. Fortunately, because there are physical and chemical restrictions on which *types* of minerals occur in equilibrium with each other, it is possible to identify certain groups of associated minerals and give each group a particular rock name (see Table 3.2).

Properties of Minerals

Minerals commonly are identified by **physical properties**. An example is hardness, which is obtained by scratching an unknown mineral with materials of known hardness. The standard reference set is **Mohs' hardness scale**, Table 2.4. (Hardness reflects the strength of the bonds between ions or atoms, thus a mineral, like kyanite, may have slightly different hardness when scratched in different directions.) Minerals that are hard and have an attractive appearance when polished are called **gemstones**. These include varieties of quartz (agate,

TABLE 2.2 Properties of the Most Common Minerals of the Earth's Crust

Mineral or group name		Varieties and chemical composition	Form, diagnostic characters	Cleavage, fracture	Color	Hardness
LIGHT COLORED MINERALS, **VERY ABUNDANT** IN EARTH'S CRUST IN ALL **MAJOR ROCK TYPES**	**FELDSPAR** (FRAMEWORK SILICATES)	*POTASSIUM FELDSPARS* Sanidine Orthoclase Microcline [KAlSi$_3$O$_8$]	Cleavable coarsely crystalline or finely granular masses. Isolated crystals or grains in rocks, most commonly not showing crystal faces.	Two at right angles, one perfect and one good; pearly luster on perfect cleavage.	White to gray, frequently pink or yellowish; some green.	6
		PLAGIOCLASE FELDSPARS Albite [NaAlSi$_3$O$_8$] Anorthite [CaAl$_2$Si$_2$O$_8$]		Two at nearly right angles; one perfect, one good. Fine parallel striations on perfect cleavage.	White to gray, less commonly greenish or yellowish.	
	QUARTZ	SiO$_2$	Single crystals or masses of 6-sided prismatic crystals. Also formless crystals and grains or finely granular or massive.	Very poor or nondetectable; conchoidal fracture.	Colorless, usually transparent; also slightly colored smoky gray, pink, yellow.	7
	MICA (SHEET SILICATES)	*MUSCOVITE* [KAl$_2$Si$_3$O$_{10}$(OH)$_2$]	Thin, disc-shaped crystals, some with hexagonal outlines. Dispersed or aggregates.	One perfect; splittable into very thin, flexible, transparent sheets.	Colorless; slight gray or green to brown in thick pieces.	2–2½
DARK COLORED MINERALS **ABUNDANT** IN MANY KINDS OF **IGNEOUS** AND **METAMORPHIC ROCKS**		*BIOTITE* [K(Mg,Fe)$_3$AlSi$_3$O$_{10}$(OH)$_2$]	Irregular, foliated masses; scaly aggregates.	One perfect; splittable into thin, flexible sheets.	Black to dark brown. Translucent to opaque.	2½–3
		CHLORITE [(Mg,Fe)$_5$(Al,Fe)$_2$Si$_3$O$_{10}$(OH)$_8$]	Foliated masses or aggregates of small scales.	One perfect: thin sheets flexible but not elastic.	Various shades of green.	2–2½
	AMPHIBOLE (DOUBLE CHAINS)	*TREMOLITE–ACTINOLITE* [Ca$_2$(Mg,Fe)$_5$Si$_8$O$_{22}$(OH)$_2$]	Long, prismatic crystals, usually 6-sided. Commonly in fibrous masses or irregular aggregates.	Two good cleavage directions at 56° and 124° angles.	Pale to deep green. Pure tremolite white, vitreous luster.	5–6
		HORNBLENDE [Complex Ca,Na,Mg,Fe, Al silicate]			Dark green to black.	

(From EARTH, Second Edition, by Frank Press and Raymond Siever. W. H. Freeman and Company. Copyright © 1978.)

TABLE 2.2 continued

Group	Mineral	Formula	Occurrence / Crystal Form	Cleavage / Fracture	Color	Hardness
SINGLE CHAINS	PYROXENE — ENSTATITE–HYPERSTHENE	$[(Mg,Fe)_2Si_2O_6]$	Prismatic crystals, either 4- or 8-sided. Granular masses and scattered grains.	Two good cleavage directions at about 90°.	Green and brown to grayish or greenish white.	5–6
	DIOPSIDE	$[(Ca,Mg)_2Si_2O_6]$			Light to dark green.	
	AUGITE	[Complex Ca,Na,Mg,Fe, Al silicate]			Very dark green to black.	
ISOLATED TETRAHEDRA	OLIVINE	$[(Mg,Fe)_2SiO_4]$	Granular masses and disseminated small grains.	Conchoidal fracture.	Olive to grayish green and brown.	6½–7
	GARNET	[Ca,Mg,Fe,Al silicate]	Isometric crystals, well-formed or rounded; high specific gravity, 3.5–4.3.	Conchoidal and irregular fracture.	Red and brown, less commonly pale colors.	6½–7½
CARBONATES	CALCITE	$CaCO_3$	Coarsely to finely crystalline in beds, veins, and other aggregates. Cleavage faces may show in coarser masses. Calcite effervesces rapidly, dolomite slowly, only in powders.	Three perfect cleavages, at oblique angles; splits to rhombohedral cleavage pieces.	Colorless, transparent to translucent; variously colored by impurities.	3
	DOLOMITE	$CaMg(CO_3)_2$				3½–4
HYDROUS ALUMINO-SILICATES	CLAY MINERALS — KAOLINITE	$[Al_2Si_2O_5(OH)_4]$	Earthy masses in soils; bedded; in association with other clays, iron oxides, or carbonates. Plastic when wet; montmorillonite swells when wet.	Earthy, irregular.	White to light gray and bl.ff; also gray to dark gray, greenish gray, and brownish depending on impurities and associated minerals.	1½–2½
	ILLITE	[similar to Muscovite + Mg,Fe]				
	MONTMORILLONITE	[Complex Ca,Na,Mg,Fe Al silicate + H_2O]				
SULFATES	GYPSUM	$CaSO_4 \cdot 2H_2O$	Granular, earthy, or finely crystalline masses. Tabular crystals.	One perfect, splitting to fairly thin slabs or sheets. Two other good cleavages.	Colorless to white. Transparent to translucent.	2
	ANHYDRITE	$CaSO_4$	Massive or crystalline aggregates in beds and veins.	One perfect, one nearly perfect, one good; at right angles.	Colorless, some tinged with blue.	3–3½
	HALITE	$NaCl$	Granular masses in beds. Some cubic crystals. Salty taste.	Three excellent cleavages at right angles.	Colorless, transparent to translucent.	2½
	OPAL–CHALCEDONY	SiO_2 [Opal is an amorphous variety; chalcedony is a formless microcrystalline quartz.]	Beds in siliceous sediments and chert; in veins or banded aggregates.	Conchoidal fracture.	Colorless or white when pure, but tinged with various colors by impurities in bands, especially in agates.	5–6½

LIGHT COLORED, TYPICALLY AS ABUNDANT CONSTITUENTS OF SEDIMENTS AND SEDIMENTARY ROCKS

TABLE 2.2 continued

Mineral or group name		Varieties and chemical composition	Form, diagnostic characters	Cleavage, fracture	Color	Hardness
DARK MINERALS COMMON IN MANY ROCK TYPES	IRON OXIDES — MAGNETITE	Fe_3O_4	Magnetic. Disseminated grains, granular masses; occasional octahedral isometric crystals. High specific gravity: 5.2.	Conchoidal or irregular fracture.	Black, metallic luster.	6
	IRON OXIDES — HEMATITE	Fe_2O_3	Earthy to dense masses, some with rounded forms, some granular or foliated. High specific gravity: 4.9–5.3.	None; uneven, sometimes splintery fracture.	Reddish-brown to black.	5½–6½
	IRON OXIDES — "LIMONITE"	$HFeO_2$ [GOETHITE is the major mineral of the mixture called "limonite," a field term.]	Earthy masses, massive bodies or encrustations, irregular layers. High specific gravity: 3.3–4.7.	One excellent in the rare crystals; usually an earthy fracture.	Yellowish-brown to dark brown and black.	5–5½
LIGHT COLORED MINERALS, MAINLY IN IGNEOUS AND METAMORPHIC ROCKS AS COMMON OR MINOR CONSTITUENTS	ALUMINOSILICATES — KYANITE	Al_2SiO_5	Long, bladed or tabular crystals or aggregates.	One perfect and one poor, parallel to length of crystals.	White to light-colored or pale blue.	5 parallel to crystal length 7 across crystals
	ALUMINOSILICATES — SILLIMANITE	Al_2SiO_5	Long, slender crystals or fibrous, felted masses.	One perfect parallel to length, not usually seen.	Colorless, gray to white.	6–7
	ALUMINOSILICATES — ANDALUSITE	Al_2SiO_5	Coarse, nearly square prismatic crystals, some with symmetrically arranged impurities.	One distinct; irregular fracture.	Red, reddish-brown, olive-green	7½
	ALKALI SILICATES — FELDSPATHOIDS	NEPHELINE [(Na.K)AlSiO₄]	Compact masses or as embedded grains, rarely as small prismatic crystals.	One distinct. Irregular fracture.	Colorless, white, light gray. Gray-greenish in masses, with greasy luster.	5½–6
	ALKALI SILICATES — FELDSPATHOIDS	LEUCITE [KAlSi₂O₆]	Trapezohedral crystals embedded in volcanic rocks.	One very imperfect.	White to gray.	5½–6
	MAGNESIUM SILICATES — SERPENTINE	$Mg_3Si_4O_{10}(OH)_8$	Fibrous (asbestos) or platy masses.	Splintery fracture.	Green; some yellowish, brownish, or gray. Waxy or greasy luster in massive habit; silky luster in fibrous habit.	4–6
	MAGNESIUM SILICATES — TALC	$Mg_3Si_4O_{10}(OH)_2$	Foliated or compact masses or aggregates.	One perfect, making thin flakes or scales. Soapy feel.	White to pale green. Pearly or greasy luster.	1

TABLE 2.2 continued

Group	Mineral	Class	Formula	Occurrence / Habit	Cleavage / Fracture	Color	Hardness
DARK COLORED MINERALS COMMON IN METAMORPHIC ROCKS	CORUNDUM	OXIDE	Al_2O_3	Some rounded, barrel-shaped crystals; most often as disseminated grains or granular (emery) masses.	Irregular fracture.	Usually brown, pink, or blue. Emery black. Gem stone varieties: ruby, sapphire.	9
	EPIDOTE	SILICATES	$Ca_2(Al,Fe)Al_2Si_3O_{12}(OH)$	Aggregates of long prismatic crystals, granular or compact masses, embedded grains.	One good, one poor at greater than right angles. Conchoidal and irregular fracture.	Green, yellow-green, gray, some varieties dark brown to black.	6–7
	STAUROLITE	SILICATES	$Fe_2Al_9Si_4O_{22}(O,OH)_2$	Short prismatic crystals, some cross-shaped, usually coarser than matrix of rock.	One poor.	Brown, reddish, or dark brown to black.	7
METALLIC LUSTER, COMMON IN MANY ROCK TYPES. ABUNDANT IN VEINS	PYRITE	SULFIDES	FeS_2	Granular masses or well-formed cubic crystals in veins and beds or disseminated. Specific gravity high: 4.9–5.2.	Uneven fracture.	Pale brass-yellow.	6–6½
	GALENA	SULFIDES	PbS	Granular masses in veins and disseminated. Some cubic crystals. Specific gravity very high: 7.3–7.6.	Three perfect cleavages at mutual right angles, giving cubic cleavage fragments.	Silver-gray.	2½
	SPHALERITE	SULFIDES	ZnS	Granular masses or compact crystalline aggregates. Specific gravity high: 3.9–4.1.	Six perfect cleavages at 60° to one another.	White to green, brown, and black. Resinous to submetallic luster.	3½–4
	CHALCOPYRITE	SULFIDES	$CuFeS_2$	Granular or compact masses; disseminated crystals. Specific gravity: 4.1–4.3.	Uneven fracture.	Brassy to golden-yellow.	3½–4
	CHALCOCITE	SULFIDES	Cu_2S	Fine-grained masses. Specific gravity: 5.5–5.8.	Conchoidal fracture.	Lead-gray to black. May tarnish green or blue.	2½–3
MINERALS, FOUND IN MINOR AMOUNTS IN A VARIETY OF ROCK TYPES AND IN VEINS OR PLACERS	RUTILE	TITANIUM OXIDES	TiO_2	Slender to prismatic crystals; granular masses. Specific gravity: 4.25.	One distinct, one less distinct. Conchoidal fracture.	Reddish-brown, some yellowish, violet, or black.	6–6½
	ILMENITE	TITANIUM OXIDES	$FeTiO_3$	Compact masses, embedded grains, detrital grains in sand. Specific gravity: 4.79.	Conchoidal fracture	Iron-black metallic to submetallic luster.	5–6
	ZEOLITES	SILICATES	Complex hydrous silicates, many varieties of minerals, including analcime, natrolite, phillipsite, heulandite, and chabazite.	Well-formed radiating crystals in cavities in volcanics, in veins, and hot springs. Also as fine-grained and earthy bedded deposits.	One perfect for most.	Colorless, white, some pinkish.	4–5

TABLE 2.3   Common Minerals in the Earth's Crust (volume %)

Feldspar	51	Plagioclase 39
Quartz	12	Orthoclase 12
Pyroxene	11	
Amphibole	5	
Micas	5	
Clay minerals	5	
Olivine	3	
Other silicates	5	
Non-silicates	4	

Data from Ronov and Yaroshevsky (1969)

TABLE 2.4   Mohs' Hardness Scale

1	Talc	(softest)
2	Gypsum	
		Fingernail
		Penny
3	Calcite	
4	Fluorite	
5	Apatite	
		Pocket Knife
		Glass
6	Orthoclase	
7	Quartz	
8	Topaz	
9	Corundum	
10	Diamond	(hardest)

amethyst, etc.), varieties of corundum (ruby and sapphire), garnet, diamond, and a host of other semiprecious and precious stones. (Several are used as industrial abrasives because of their hardness.)

Other useful physical properties include **cleavage** (definite, repeated planar breaks, reflecting lesser bond strengths), **streak** (color of the mineral powder left after scratching a porcelain plate), and **luster, color,** etc., as listed in Table 2.2.

Minerals can show markedly different physical properties (resulting from different atomic lattice structures) while having the same chemical composition. This phenomenon is called **polymorphism** and is well illustrated by carbon, which occurs as the hardest known material, diamond, and also as the soft lubricant, graphite. Conversely, minerals may show **isomorphism,** the same physical appearance but different chemical composition. If one element takes the place of another so the composition varies within certain limits, for example, the plagioclase feldspars, which range from calcium-rich to sodium-rich, or olivine which ranges from magnesium- to iron-rich, the phenomenon is called **solid solution.** Many isomorphous minerals show solid solution and vice-versa, but each property can exist independently of the other.

Silicate minerals occur in one of six **structural groups,** reflecting the different arrangements of the "building block" which is a silica tetrahedron (one silicon surrounded by four oxygens) shown in Figs. 2.2 and 2.3. These groups have different physical properties (e.g., cleavage) and different temperature/pressure ranges over which they are stable. The structural groups are listed in Table 2.5.

It is found that the 2200-odd minerals can be assigned to one of six **crystal systems** defined by the length of reference axes and the angles at which they meet. For example, the *isometric* or cubic system has three equal axes (a, b, and c, in Fig. 2.4) that meet at right angles. Note that even if the edges and corners of the cube were removed, the resultant form would still be isometric (although no longer a cube).

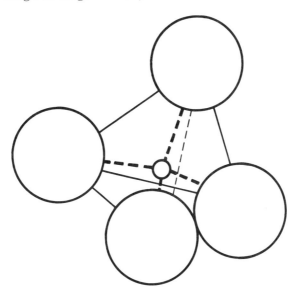

Figure 2.2. The silica tetrahedron. Four large oxygen ions surround a smaller silicon ion, forming a tetrahedron. Dashed lines show bonds between silicon and oxygen ions.

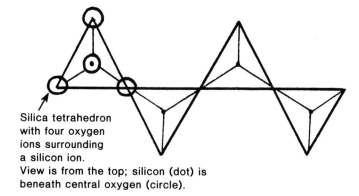

Silica tetrahedron
with four oxygen
ions surrounding
a silicon ion.
View is from the top; silicon (dot) is
beneath central oxygen (circle).

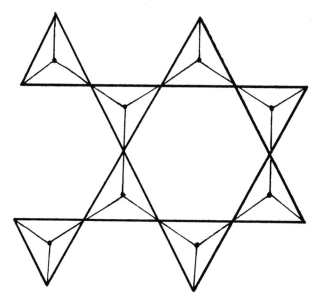

Figure 2.3. Diagram showing the arrangement of silica
tetrahedra in single chain and double chain minerals (sheet
structure would extend the double chain in the plane of the
page).

TABLE 2.5   Major Silicate Structures

Arrangement of SiO$_4$ tetrahedra	Si/O ratio	Mineral example
Isolated tetrahedra linked by cations	1:4	Olivine
Rings of tetrahedra	1:3	Tourmaline
Single chains; chains bonded together by cations	1:3	Pyroxenes
Double chains	4:11	Amphiboles
Sheets bonded together by cations or alumina sheets	2:5	Micas Clays
Three-dimensional frameworks; all oxygens shared with other tetrahedra.	1:2	Feldspar Quartz

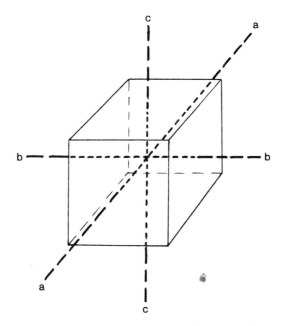

Figure 2.4. Mutually-perpendicular axes of equal length defining the isometric (cubic) crystal system.

If the c axis were lengthened, the mineral would belong to a different crystal system, the *tetragonal.* The other systems represent further similar modifications (*orthorhombic, monoclinic, triclinic, hexagonal*). Not surprisingly, since external form reflects the internal arrangement of atoms, different minerals crystallizing in the same system have many similar physical properties.

Minerals and Absolute Age

In 1896 Henri Becquerel discovered **radioactivity.** Kelvin's estimates of the age of the earth, which were based on cooling rates, became untenable as a source of thermal energy was identified for the earth; the contracting-earth hypothesis (p. 208) was dealt a severe blow. Further, the phenomenon of radioactivity provided a method for dating minerals and rocks to give an **absolute age** (in years) where only relative age determinations had been possible (Table 4.4).

All elements have **isotopes** (same number of protons—the atomic number, which defines the element—but different numbers of neutrons). Some isotopes are unstable and undergo radioactive decay whereby alpha, beta and other particles are given off from the nucleus. As Einstein showed, $e = mc^2$, thus the energy (e) released is equal to the mass of the particle (m) times the square of the speed of light; this enormous amount of energy is harnessed in nuclear power plants, released in atom bombs, and gives much of the earth's internal heat through natural decay.

The rate at which a radioactive element (parent) will decay to another element (daughter) is independent of temperature, pressure, or chemical reactions. The time for one unit of parent element to decay to half that amount (plus daughter product) is called the **half-life** (Fig. 2.5). Obviously, as decay proceeds, the ratio of parent to daughter changes with time. Thus if we measured that proportion we could, knowing the rate of decay, determine how long the change had been taking place. Since the elements are part of minerals, and the minerals constitute rocks, we can establish the age of a rock and the geologic events of which it is a part.

In igneous rocks the age is of the formation of minerals crystallizing out of the magma. For detrital sedimentary rocks, the age would be of materials at the source area that were eroded to give the sediment (erosion of Precambrian shields, for example, would give mineral fragments perhaps one or two billion years old, but they could form present-day beach sands). Radiometric age-dating of these minerals would not tell us the age of the beach, or of the sedimentary rock formed if the sediment became compacted and cemented.

Metamorphic rocks yield ages that indicate the time of metamorphism. Recrystallization essentially re-sets the decay clock by redistributing the isotopes into newly-formed minerals. Some systems, such as uranium-lead, also may give the age of the *original* material.

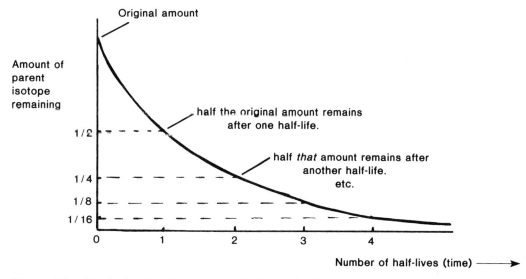

Figure 2.5. Graph showing the exponential relationship between the amount of parent isotope and time (as radioactive decay proceeds).

Major decay systems used include:

a. **Uranium-Lead**

$U^{238} \rightarrow Pb^{206}$ Half-life: 4.5×10^9 years (The numbers 238 and 206 indicate the number of protons plus neutrons in the nucleus.)

$U^{235} \rightarrow Pb^{207}$ Half-life: 0.71×10^9 years

Effective dating range: 10^7 to 4.7×10^9 years (age of Earth).
Minerals dated: zircon, pitchblende, uraninite

b. **Rubidium-Strontium**

$Rb^{87} \rightarrow Sr^{87}$ Half-life: 47×10^9 years

Effective dating range: 50×10^7 to 4.7×10^9 years
Minerals dated: micas, feldspars, whole rocks.

Potassium-Argon

$K^{40} \rightarrow Ar^{40}$ Half-life: 1.3×10^9 years

Effective dating range: 30×10^3 to 4.7×10^9 years
Minerals dated: micas, amphiboles, feldspars, whole rock. (Argon is an inert gas; this system is disturbed easily as heating drives off some of the gas. This loss will decrease the amount of daughter present, thus giving a measured age that is younger than the real age.)

Carbon 14

$C^{14} \rightarrow N^{14}$ Half-life: $5{,}730 \pm 30$ years

Effective dating range: 0 to 50×10^3 years (especially less than 20×10^3).

Minerals dated: carbon-containing minerals and materials.

Carbon-14 is produced from nitrogen-14 by cosmic-ray bombardment in the upper atmosphere; it decays back to nitrogen-14. The amount of C-14 present in a living organism remains essentially constant. If a carbon-containing compound is buried and removed from equilibration with the atmosphere, the decaying C-14 is not replenished with new C-14; the longer it is buried, the less C-14 will still be present. The amount of C-14 relative to other carbon isotopes is the basis of age dating.

Rocks

The solid earth consists of minerals aggregated with other minerals to form igneous, sedimentary or metamorphic rocks, as well as regolith.

It should become clear in the following sections that interpretation of a rock (origin, age, history, etc.), depends heavily on an interpretation of the minerals making up that rock (their composition and mutual arrangement—size, shape, orientation, etc., illustrated in Figure 2.1) as outlined on page 24. If minerals may be thought of as the "words" in the geologic record (the elements making up the mineral would be the "letters"), then rocks are the "sentences". (Words do exist and have importance as separate entities but their significance is greater when they are combined into sentences.) Similarly, as sentences convey more information when placed together, so a sequence of rock layers or successive intrusions can give additional information on the geologic history of an area. The study of rocks is called *petrology*.

3. Igneous Rocks

Igneous rocks are very closely related to the tectonic cycle and form by the cooling of *magma* which is generated by partial melting either deep in the crust (granitic magmas) or in the deeper lithosphere or asthenosphere (basaltic magmas). Basaltic (Table 3.2) magmas form at temperatures over 1000°C; granitic magmas at probably several hundred degrees less. Controls on the generation of magma include heat from radioactive decay, or lowering the melting temperature through an increase in water content or a decrease in confining pressure. During plate collision (Figs. 3.17 and 3.18), water-saturated crustal rocks and sediment, of generally low melting temperatures, are subducted into higher temperature parts of the mantle where partial melting occurs. At a spreading ridge (Fig. 3.19), high-temperature mantle material is brought into a lower pressure environment and magma is produced. If magma reaches the surface it is called lava and **extrusive** (volcanic) igneous rocks result (Fig. 3.1). If magma crystallizes below the earth's surface, then **intrusive** rocks are formed. Major intrusive bodies include: Sheet-like **dikes** (cross-cutting or discordant) shown in Fig. 3.2, and **sills** (concordant) that are generally basaltic in composition (a domed sill, usually of granitic composition, is called a **laccolith**); saucer- or funnel-shaped masses of basaltic (gabbroic) composition form **lopoliths,** typically several hundred kilometers in diameter; irregularly-shaped, large masses of granitic composition are called **batholiths.** (Note that we see intrusive bodies only because the overlying rocks have been removed by erosion.)

Figure 3.1. Lava flow advancing down a small road, Hawaii. (Photo, O. Don Hermes.)

Figure 3.2. Dike exposed by erosion. The dike is more resistant than the surrounding rocks thus it forms a ridge. Shiprock, New Mexico. (Photo, Jon C. Boothroyd.)

The Coast Range Batholith of British Columbia is over 2000 km long and up to 300 km wide; the Sierra Nevada Batholith of California is 650 km by over 75 km. Major batholiths represent multiple intrusions during magmatic episodes extending over long periods (8-10 million years for the Boulder [Montana] Batholith, 120-150 million years for the Sierra Nevada). Batholiths occur in the cores of folded mountain ranges (Fig. 1.3).

The forms of some igneous intrusive bodies and volcanic features are shown diagrammatically in Fig. 3.3.

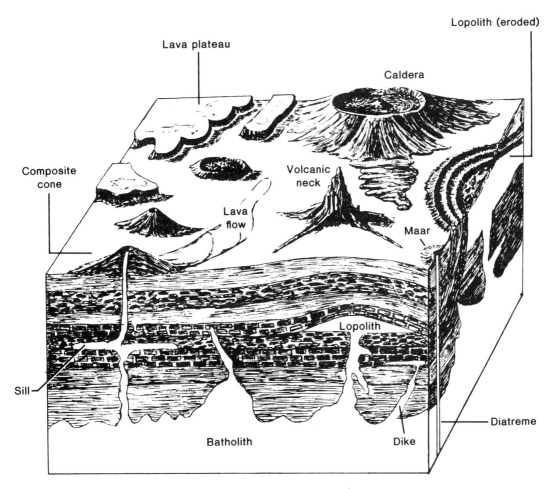

Figure 3.3 Idealized block diagram showing some intrusive and extrusive igneous forms. (Based on a diagram by R. G. Schmidt and H. R. Shaw, U. S. Geological Survey.)

Volcanic Activity

Basaltic magma dominates the ocean basins (ridges, floor and islands) and occurs as flood (plateau) basalts in tensional environments on continents (the Columbia River basalts, Fig. 3.4, cover more than 120,000 square kilometers). Andesitic and rhyolitic (Table 3.2) magmas, as well as basaltic, occur where plates collide along continental margins or curved belts of volcanic islands called island arcs (Fig. 3.5). Figure 3.6 gives the location of active volcanoes. The *Andesite Line* follows the western island arcs and eastern continental margins of the Pacific Ocean; inside this area andesites do not occur. Figure 3.7 shows the distribution of recent major volcanic eruptions (see also, Fig. 1.3).

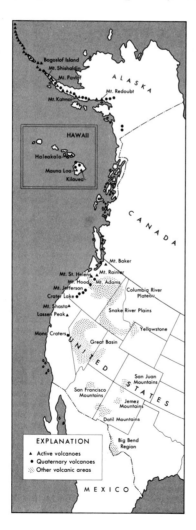

Figure 3.4. Recent volcanic activity in the United States. (From U. S. Geological Survey.)

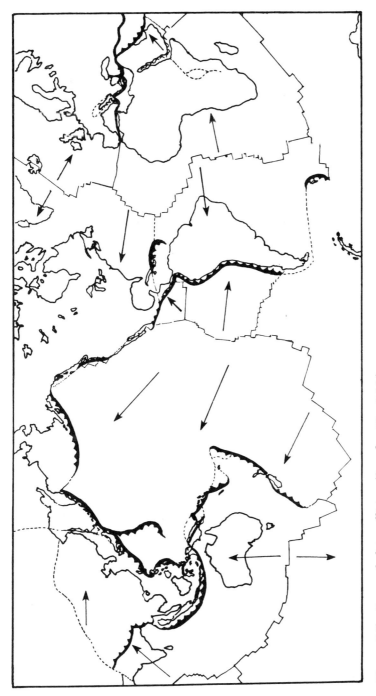

Figure 3.5. Zones of plate collision. (See also, Fig. 1.11.)

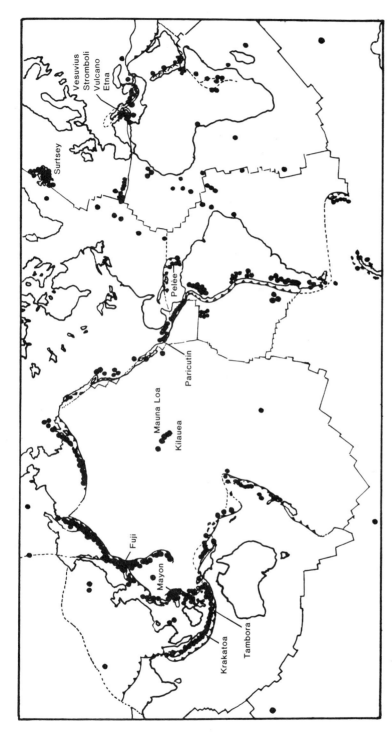

Figure 3.6. Active volcanoes of the world. Fewer than 30 of the 500 or so active volcanoes erupt in a given year. Note relationship to plate boundaries, especially subduction zones. The circum-Pacific belt is called the Ring of Fire.

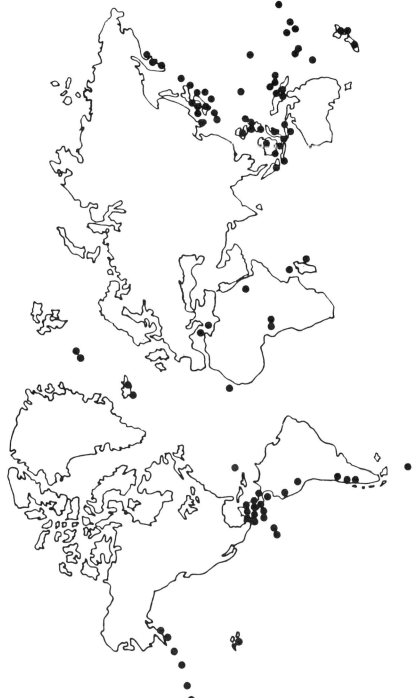

Figure 3.7. Major volcanic eruptions 1968-1974 (cf. Fig. 17.1). (Reported by the Center for Short-Lived Phenomena.)

Volcanic gases contribute to the development of the atmosphere and hydrosphere (the original atmosphere and hydrosphere developed during the early degassing of the earth). Volcanic activity also can modify the climate by ejecting large volumes of ash into the upper atmosphere, causing a reduction in the solar radiation received on earth; some have suggested that the onset of glaciation may be related to periods of particularly active volcanism.

Volcanic activity probably is responsible for the legend of Atlantis. About 1500 B.C. a violent eruption of the volcano Santorin in the Aegean Sea exploded most of what is now called Thera and the prosperous Minoan civilization on Crete, 120 km away, was largely destroyed by ash fall and sea waves. Around 380 B.C., Plato created the story of Atlantis from ancient Greek and Egyptian legends about a "lost island".

Types of Volcanic Eruptions

Most eruptions are from the summit of a central vent but flank eruptions (from the sides of the volcano) also are common. Continental volcanoes typically have a pipe-like vent; oceanic volcanic activity commonly is from fissures.

The shape of a volcano, the violence of its eruptions, the type of material ejected (lava; pyroclastic, Table 3.1) and the composition of that material are all interrelated. Fundamental controls are the magma's viscosity and volatile content. A given volcano may exhibit a wide range of eruptive styles and products during its history.

1. *Relatively quiet eruptions*

   Fluid magma (low viscosity)
   Non-explosive (gases escape easily because of low viscosity)
   Largely lava, minor pyroclastics
   Typically basalts (low silica and high iron content, hence, low viscosity)
   Extruded as:
   a. widespread sheets of lava (*flood, plateau,* or *fissure flows* such as the Columbia River basalts)
   b. **shield volcanoes**—low, broad edifices; so-called **Hawaiian-type** eruption.

2. *Explosive eruptions*

   General increase in eruptive violence and pyroclastics as follows:
   a. **Strombolian-type** (Stromboli, Italy; Paricutin, Mexico)
   Less fluid (more viscous) magma than above, forms thicker, less widespread flows
   More pyroclastics—bombs, blocks, lapilli, ash (Fig. 3.8)
   Basaltic to andesitic composition
   Conical shape (Fig. 3.9), commonly with alternating layers of lava and pyroclastics (stratovolcano or composite cone); cinder cones from brief eruptions.

Figure 3.8. Volcanic eruption, Mount Etna, Sicily, showing paths of ejected bombs and blocks. (Photo, Eugene J. Tynan.)

   b. **Vulcanian-type** (Vulcano and Vulcanello, Italy (Fig. 3.10)
      Viscous lava; mostly pyroclastics—ash to blocks
      Ash and cinder cones
      Andesitic to rhyolitic composition.
   c. **Pelean-(or Peléean) type** (Mt. Pelée, Martinique; Mayon, Philippines).
      Type of Vulcanian eruption with glowing avalanches (nueés ardentes) of pyroclastics,
         giving welded tuffs
      Steep dome of viscous lava
      Andesitic to rhyolitic composition.
   d. **Plinian-type** (Vesuvius eruption of 79 A.D.; Krakatoa, Indonesia)
      Sheets of ash and pumice (Fig. 3.11)
      Very violent form of Vulcanian eruption
      Volcano collapses to form a caldera (Fig. 3.12).

Figure 3.9A. Stromboli, Italy. (Photo, Eugene J. Tynan.)

Figure 3.9B. Sunset Crater, Arizona. (Photo courtesy of U. S. Department of the Interior, National Park Service.)

Figure 3.10. Erosion has exposed some of the layers typical of a volcanic cone. Vulcanello, Italy. (Photo, Eugene J. Tynan.)

Figure 3.11. Volcanic ash layers near Rome, Italy. (Photo, O. Don Hermes.)

Figure 3.12. Crater Lake, Oregon. The lake occupies a caldera formed when the summit of a large volcano collapsed. The island is a small, younger volcano. (Photo courtesy U. S. Department of the Interior, National Park Service.)

TABLE 3.1   Types of pyroclastic material (tephra)

Size	Name	Rock Name
<.25 mm diameter	Volcanic dust	Tuff, welded tuff
.25-4 mm	Volcanic ash	
4-64 mm	Lapilli	Lapilli agglomerate, lapilli breccia
>64 mm	Bomb (round to subangular)	Agglomerate
	Block (angular)	Volcanic breccia

3. *Other types of volcanic activity*
   a. **Diatreme**—pipes of pyroclastic material from subsurface explosive emplacement
   b. **Ash flow sheets**—fissure or vent eruptions of rhyolitic ash and pumice.
   c. **Cinder cone**—pile of pyroclastic material, often as a satellite on the flanks of another type of volcano
   d. **Maar**—crater in non-volcanic (with perhaps some pyroclastic) material. Possibly on top of a diatreme (Fig. 3.3).
   e. **Fumarole**—vent emitting nothing but gas. Gases are predominantly steam, together with sulfur and sulfur oxides, carbon dioxide and monoxide, hydrochloric and hydrofluoric acid, hydrogen, nitrogen, etc.
   f. **Spatter cone**—congealed lava deposited directly around a relatively quiet vent.
   g. **Hot springs** (Fig. 10.10), **geysers** and **mud volcanoes**—vents issuing water and/or stream (heated groundwater, a source of geothermal energy).

If lava has a very viscous crust that is fragmented by movement of the flow, a blocky or clinkery surface is formed, called **aa**. If the lava has more dissolved gas (lower viscosity) the crust may remain plastic and gas bubbles may form *vesicles* in the lava underneath; this smooth or ropy-surfaced lava is called **pahoehoe** (Fig. 3.13). When lava cools slowly in the interior of a flow, it commonly forms *columnar joints* (Fig. 3.14).

Predicting an Eruption

It is possible in some cases to predict volcanic eruptions. Prediction is important not only because of the obvious direct effects of an eruption, but also because of indirect effects, such as mudflows of volcanic ash (lahars), and floods (melting mountain glaciers, sea waves from submarine or ocean island explosive eruptions). Indicators of an eruption include a) an increase in the number of small earthquake tremors, b) upward tilt of the area around the vol-

Figure 3.13. Pahoehoe lava flow showing characteristic ropy surface. Vesuvius. (Photo, Eugene J. Tynan.)

Figure 3.14. Columnar jointing shown by two lava flows. (Lavas are separated by Pleistocene gravel deposits.) Yellowstone National Park. (Photo, Jon C. Boothroyd.)

cano, c) increase in the temperature of volcanic gases as well as a change in composition, d) decrease in the strength of the local magnetic field. These effects are caused by the movement of magma into a near-surface position.

Crystallization of Magma

Minerals crystallize from a magma in a predictable sequence known as **Bowen's Reaction Series,** given in Fig. 3.15. A completely liquid magma of basaltic composition, when cooled, first crystallizes olivine and calcium-rich plagioclase. As the temperature of the magma continues to fall, olivine reacts with the liquid and dissolves back into the magma; pyroxene then crystallizes as it is the mineral stable at that lower temperature. (Plagioclase also reacts with the liquid to release some calcium and take up some sodium; sodium-rich plagioclase is stable at lower temperatures). Further decrease in temperature causes the pyroxene to react with the liquid to give amphibole which, in turn, changes to biotite at still lower tempera-

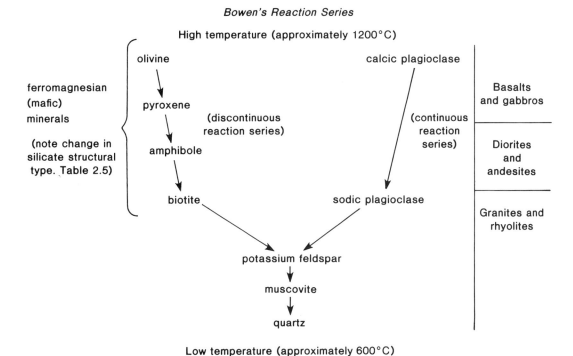

Figure 3.15. Bowen's Reaction Series, showing the sequence of minerals (and rocks) produced by fractional crystallization and differentiation of a basalt magma: If early-formed (basaltic) minerals can not react with the cooling magma (isolated by gravity settling, for example), the magma becomes enriched in elements forming the lower-temperature (granitic) minerals. (The last minerals to crystallize are the first to melt from a solid during partial melting.)

ture. The sequence olivine-to-pyroxene-to-amphibole-to-biotite is called a **discontinuous reaction series** because the four minerals are structurally and chemically distinct (Table 2.5). Plagioclase constitutes a **continuous reaction series** because the chemical changes take place within a single mineral species (this illustrates isomorphism and solid solution, page 31).

Biotite and sodic plagioclase remain stable as the temperature falls, and are joined by potassium feldspar, muscovite and quartz. (If the late-stage magma is rich in volatiles (water, etc.) these lower-temperature minerals may grow very large in the low-viscosity liquid; intrusions of this type of magma form **pegmatites**, Fig. 3.16).

This crystallization sequence is somewhat oversimplified but it illustrates a relationship between the two most common igneous rock types, basalt and granite. Basalt consists of pyroxene and calcium-rich plagioclase (olivine is present with pyroxene because slow rates of reaction in viscous fluids like magmas commonly do not allow the change to be complete before solidification occurs). Because these minerals are at the high-temperature end of the series, they would be expected in rocks that cooled too rapidly to allow further progression down the sequence. Basalt is thus the most common volcanic (extrusive) rock. (Basaltic rocks are rich in Mg and Fe and are sometimes called **mafic** rocks.) Conversely, prolonged slow cooling of the magma, aided by the removal of early minerals that sink to the bottom of

Figure 3.16. Pegmatites (white) in the Black Canyon of the Gunnison, Colorado. (Photo, O. Don Hermes)

the magma chamber (*gravity settling*) or are separated from the liquid by squeezing (*filter pressing*), should favor the assemblage quartz, muscovite, potassium-rich feldspar, sodium-rich plagioclase and biotite; these silica-rich minerals form so-called **felsic** rocks, typified by granite, the most common intrusive igneous rock (20% of the volume of the earth's crust). (Note, however, that most granites result from the cooling of a magma that is already granitic in composition, produced by partial melting of continental crust during plate collision.) Granites also may originate by metamorphic or metasomatic processes (essentially in the solid state with perhaps some intergranular fluid) whereby pre-existing rocks (for example, sandstone) are changed into granite by migration of ions which produce new minerals and recrystallize others, without large-scale melting to form magma. When metamorphism produces granitic rocks the process is called **granitization.**

In addition to differences in mineralogy between granite and basalt, there are differences in texture (size, shape, and arrangement of minerals) so that basalts, cooled quickly, typically have small crystals (or even some glass where cooling is extremely rapid) and granite, cooled slowly, has large crystals (Fig 2.1). It has been estimated that a gabbro sill some 200 meters thick would crystallize in about 700 years; large batholiths may take up to 10 million years to solidify.

Classification of Igneous Rocks

The two aspects, mineralogy and texture, are incorporated into a classification scheme for the naming of igneous rocks, shown in Table 3.2.

Economic Significance

Most of the world's metallic mineral resources occur as ore minerals (mainly sulfides or oxides) associated with igneous rocks. Most economically important deposits are found adjacent to granitic-type igneous masses (Fig. 6.1) and represent precipitation from high-temperature fluids expelled from the cooling magma into the surrounding rocks (so-called **metasomatic** and **hydrothermal** processes). Certain other metals are found in pegmatites (e.g., lithium), or *within* large igneous masses (usually gabbros or peridotites), for example, Pt, Cr, Ni, and the well-known non-metal, diamond (see also, Section 6, Mineral Resources, especially Table 6.1).

Relationship to Plate Boundaries

The formation of batholiths and volcanoes in a **convergent plate boundary** (subduction zone), such as the Andes, is illustrated in Figure 3.17. Figure 3.18 shows a similar tectonic setting but without a thick continental crust. This is an **island-arc system** like the West Indies, Java, the Philippines, Japan, etc. Explosive volcanic activity is associated with this tectonic environment (silicia-rich magmas such as andesites and, especially, rhyolites are more viscous and tend to be more explosive than silica-poor basalts). Famous, or infamous, volcanoes in island arcs include Mont Pelée, Krakatoa, and Tambora (Fig. 3.6).

TABLE 3.2 Igneous rock classification

As illustrated in Bowen's Reaction Series, mineral assemblages change as the temperature of the magma falls. Also, the rate of cooling, content of volatiles (water, etc.) influence the size of minerals crystallizing from the magma. These two aspects, mineral composition and texture, are incorporated in the classification scheme:

MINERAL COMPOSITION

Minerals:
- Quartz
- Potassium feldspar
- Plagioclase (sodium-rich —————— calcium-rich)
- Ferro-magnesian minerals:
  - Muscovite
  - biotite
  - amphibole
  - pyroxene
  - olivine

Proportions (Feldspar / Quartz / Ferromagnesians): 75% — 50% — 25%

ROCKS	(Felsic)		(Intermediate)	(Mafic)	(Ultramafic)
Coarse-grained* (Intrusive)	Granite	Granodiorite	Diorite	Gabbro	Peridotite
Fine-grained (Extrusive)	Rhyolite	Dacite	Andesite	Basalt	No common volcanic equivalent
Glassy (Extrusive)	Obsidian (massive) Pumice (frothy)				
Fragmental (Extrusive) (Pyroclastic)	Tuff, volcanic breccia welded tuff				

*Individual minerals visible without a magnifying lens. Rocks with some minerals markedly larger than others have porphyritic texture (e.g., porphyritic andesite or porphyritic granite, shown in Fig. 2.1A).

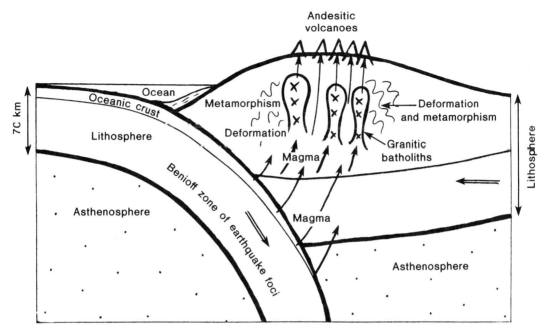

Figure 3.17. Convergent plate boundary (subduction zone) showing development of batholiths and volcanic activity in the continental crust. The major zone of earthquake foci (Benioff zone) is also shown.

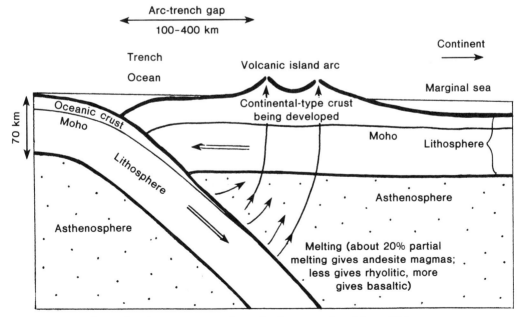

Figure 3.18. Generalized cross-section of an island arc, produced by collision between oceanic plates.

Figure 3.19 illustrates a quiet, fissure-type **spreading ocean ridge** which produces a very consistent type of basalt. The sequence of rocks making up the mafic oceanic crust and part of the upper mantle (ultramafic peridotite) is called an **ophiolite suite.** The recognition of these suites squeezed upward (obducted) within narrow zones in continental areas is taken as evidence of old plate collisions (or sutures).

It will be seen from Fig. 3.6 that some well-known volcanoes are not at plate boundaries. Hawaii is one obvious example. An explanation of activity *within* a plate involves **plumes** or **hot spots,** whereby hot mantle material which rises in a narrow plume, perhaps from the core-mantle boundary, reaches the lithosphere in the interior of a plate. Presumably, the plume would generate magma for a volcanic cycle. As the overlying plate is moving, a stationary plume should produce a line of volcanoes on the plate. Those volcanoes now farthest away from the plume should have been extinct the longest. The Hawaiian volcanic chain and Emperor seamounts show this predicted age sequence as well as a rotation or shift in spreading direction, indicated by the bend near Midway Island (Fig. 3.20). These and similar earth-quake-free oceanic chains are called **aseismic ridges.**

The collision of two **continent-bearing plates** does not produce volcanic igneous activity but granitic emplacement and metamorphism occur at depth. Continental crust is less dense than oceanic crust and mantle material, thus it tends to rise instead of being carried down the

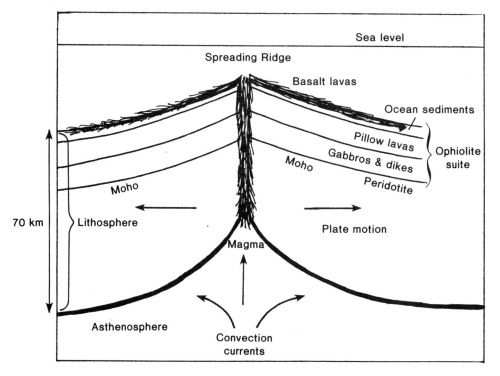

Figure 3.19. Simplified cross-section of a divergent plate boundary (spreading ridge) showing the addition of new basaltic lithosphere.

subduction zone. (Some oceanic lithosphere may get caught between the plates giving ophiolite material within the granitic crust, Fig. 3.21.) The crust from one plate is pushed against and on top of crust from the other, giving major mountain ranges such as the Himalayas.

Table 1.1 summarizes the characteristics of plate boundaries.

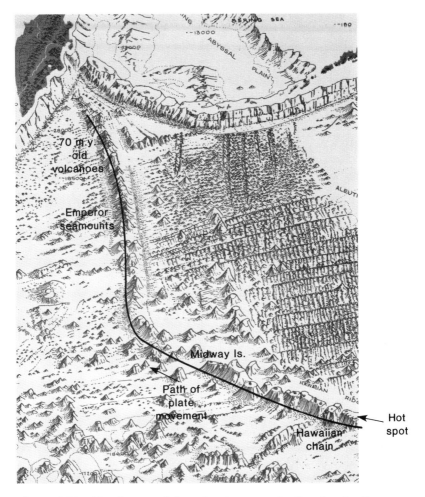

Figure 3.20. Sketch map of the volcanic mountains forming the Emperor Seamounts and Hawaiian Islands. The pattern could reflect movement of the Pacific plate over a mantle hot-spot, or plume, located under Hawaii. (Base map, courtesy Geological Society of America.)

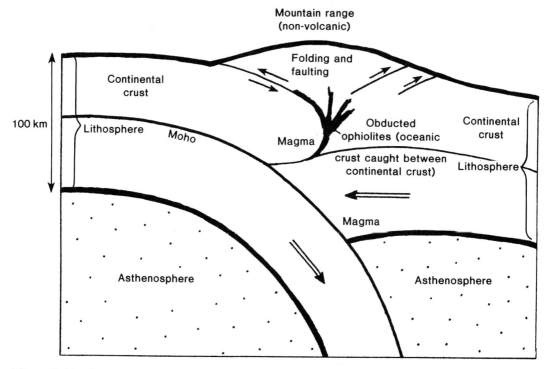

Figure 3.21. Cross-section of a convergent plate boundary involving two continent-bearing plates. Some slices of old oceanic crust (ophiolites) may get squeezed up (obducted) during collision.

4. Sedimentary Rocks

Sedimentary rocks form at the surface of the earth and are mainly products of the hydrologic cycle (Fig. 1.9).

To understand the origin of sedimentary rocks one must know something of **weathering** —the response of mineral assemblages to atmospheric temperatures, pressures and fluids (rain is a weak solution of carbonic acid). When minerals are out of chemical equilibrium with their environment they tend to break down, through oxidation, hydration, solution or other processes (see Table 7.1) so that, for example, plagioclase in an igneous rock would respond to weathering by changing, through a process called hydrolysis, to a clay mineral that is stable under near-surface conditions:

$$2Na\,Al\,Si_3O_8 + H_2CO_3 + H_2O \rightarrow$$
plagioclase feldspar + carbonic acid + water

$$Al_2\,Si_2O_5\,(OH)_4 \quad + \quad Na_2\,CO_3 \quad + 4\,SiO_2$$
kaolin (a clay mineral) + soluble sodium carbonate + silica

Note that the minerals most resistant to weathering are those at the bottom of Bowen's Series (Fig. 3.15), while those most easily affected occur at the top of the series (highest temperature of formation, thus farthest out of equilibrium with weathering conditions; different structural types, given in Table 2.5).

The products of weathering are fragments of pre-existing minerals (e.g., quartz), new minerals produced by chemical weathering (e.g., clays), and material carried away in solution (e.g., potassium, sodium, calcium). Upon deposition and compaction these materials give the two classes of sedimentary rocks; *detrital* (fragments) and *chemical* (precipitates and/or organic material).

Classification of Sedimentary Rocks

It is convenient to classify detrital sedimentary rocks primarily on the basis of grain size (Table 4.1), as this is measured easily and indicates the energy levels of the transportation process (such as by rivers or wind), as well as at the depositional site (beach, lake, ocean shelf); the large, heavier grains obviously require more energy to move them. Chemical sedimentary rocks are named mainly by composition (mineralogy). Table 4.2 is a classification scheme for sedimentary rocks.

TABLE 4.1   Wentworth Scale of Particle Sizes for Clastic Sediments

Wentworth Scale		Approximate Equivalent
Size	Fragment	
	Boulder	
256 mm		10 in.
	Cobble	
64 mm		2½ in.
	Pebble	
4 mm		5/32 in.
	Granule	
2 mm		5/64 in.
	Sand	
1/16 mm		.0025 in.
	Silt	
1/256 mm	——— Dust	.0015 in.
	Clay	

Properties such as degree of grain roundness, sorting (whether the grains are approximately the same size) and composition of fragments (including fossils) are also important in interpreting the origin of sedimentary rocks and may be used to modify the rock name. "Mature" rocks consist of rounded, well-sorted and stable minerals (e.g., quartz sandstone, not graywacke).

Origin of Sedimentary Rocks

Fig. 4.1 illustrates some of the factors in the origin of sedimentary rocks. The type of sedimentary rock formed is influenced by the sort of rock being weathered at the source area, the weathering processes, the nature of the transporting (erosional) agent, and changes produced by conditions at the site of deposition (**diagenesis**). Diagenetic changes are summarized in Fig. 4.2. Typical depositional environments are shown in Fig. 4.3. The types of sediments to be found in these environments are given in Table 4.3.

The gross distribution of sediments to be expected from continent to deep ocean basin is given in a generalized diagram as Fig. 4.4. Because different types of sedimentary material are deposited in different environments, a single sedimentary unit, deposited over a given time interval, may show lateral changes in composition, grain size, fossil content, etc. For example, the shelf sequence in Fig. 4.4 shows conglomerates near the continental source grading seaward through sandstones, siltstones, shales, and carbonates. These lateral changes in adjacent materials deposited at the same time are called **facies changes** and recognition of them is extremely important in correlating rock units (see p. 69).

The composition of sedimentary rocks is influenced by **tectonic activity.** Rapid erosion of uplifted mountain ranges yields abundant feldspar from the granitic and metamorphic rocks making up the folded range; feldspar is the prime ingredient of the sedimentary rock,

TABLE 4.2 Sedimentary Rock Classification

Sedimentary rocks produced from the cemented fragments of pre-existing rocks and minerals are called clastic, detrital or fragmental. They are classified mainly on the basis of the size of the particles. The second class of sedimentary rocks is chemical or biochemical in origin; composition is the main basis of classification.

a) Detrital Sedimentary Rocks:

Particle Size	Sediment name	Rock produced by lithification of that sediment
over 2 mm	Gravel	Conglomerate (rounded fragments), breccia (angular fragments). These rocks usually have a matrix of finer material (sand-or silt-sized).
1/16-2 mm	Sand	Sandstone, subdivided on the basis of composition: 1/ Quartzose sandstone (quartz arenite) 2/ Arkose 3/ Graywacke quartz 1 2 3 feldspar rock fragments / clay
1/256-1/16 mm	Silt	Siltstone (Loess is a wind-derived type having a buff color and conchoidal fracture)
Less than 1/256 mm	Clay	Mudstone or claystone, if massive; Shale if it shows closely-spaced bedding planes (fissility)

TABLE 4.2 continued

b) Chemical Sedimentary Rocks:

Chemical Composition	Mineral	Rock

$CaCO_3$	Calcite
	
	Aragonite
	(in young material;
	also in pearls)
		Limestone
		
		varieties include:
		abundant shell fragments cemented together: coquina (Fig. 2.1Ad)
		microscopic plants
		    and animals: chalk (deep-sea oozes)
		spongy hot or cold
		    spring deposits: tufa and travertine
		muddy limestone: marl
		cemented spheroids: oolitic limestone
		dripstone in caves: stalactites and stalagmites

$CaMg(CO_3)_2$	Dolomite	Dolostone (also called dolomite for rock as well as mineral)
$CaSO_4 \cdot H_2O$	Gypsum	Rock gypsum or gyprock
$NaCl$	Halite	Rock salt
C(plus O_2, H, etc).	not a mineral- plant remains	Coal varieties include: lignite subbituminous bituminous (based on heat output and carbon content)
Fe_2O_3, $FeO(OH)$	Hematite, Limonite	Iron formation
SiO_2	Quartz and chalcedony; varieties: opal (hydrous) agate (banded) jasper (red) flint (black)	Chert varieties: remains of microscopic plants: diatomite microscopic animals: radiolarite hot springs: siliceous sinter

Examples of evaporites

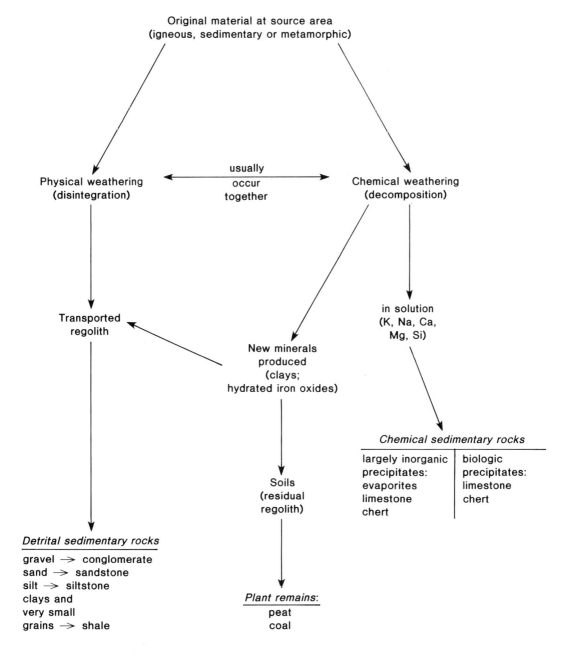

Figure 4.1.   An outline of the origins of sedimentary rocks.

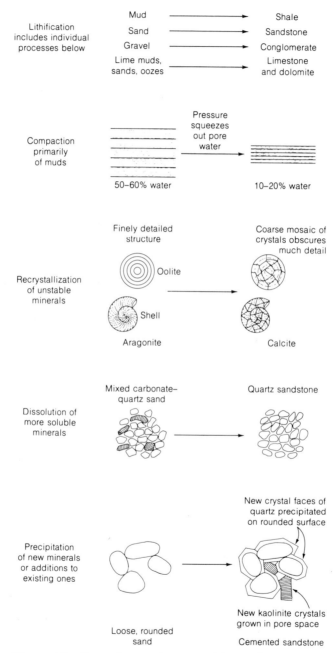

Figure 4.2. Some changes in composition and texture produced by diagenetic processes. Most of the changes transform (lithify) loose sediment into rigid sedimentary rock. (From *Earth*, Second Edition, by Frank Press and Raymond Siever. W. H. Freeman and Company. Copyright 1978.)

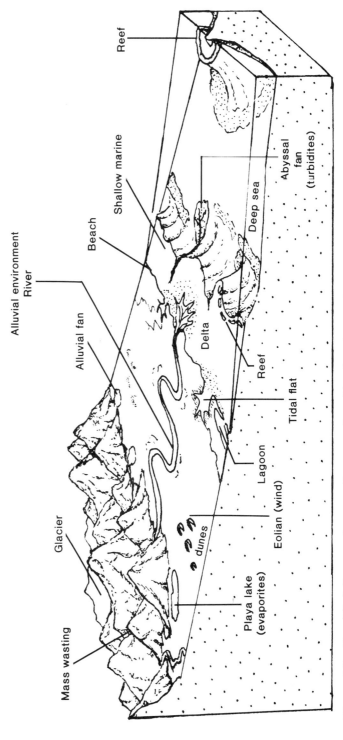

Figure 4.3. Simplified diagram of some sedimentary depositional environments.

TABLE 4.3   **Major sedimentary environments**

Environment	Sediment types
Dominantly detrital	
Alluvial	Moderate to poorly sorted point bar and channel sands and gravels interbedded with sandy shales; occasional coal or peat beds; very coarse debris in alluvial fans.
Deltaic	Well-sorted bar-finger sands and delta-front silts and clays interbedded with marine clays and coastal-swamp deposits.
Tidal	Moderately sorted tidal-channel sands, silts, and muds; abundant invertebrate faunas; pyrite in black deoxygenated layers below surface in some places.
Beach and bar	Well-sorted sands and occasional gravels with well-rounded and worn grains interbedded with marine and nonmarine deposits.
Shallow-water marine (continental shelf)	Sands, silts, and muds with variable marine fauna distributed in sheet-like geometry and interbedded frequently with some carbonate sediment.
Deep-sea turbidite (continental rise, abyssal plain, and trench)	Graded sequences of poorly sorted sand to fine clay-size particles interbedded with pelagic clays; some individual units can be traced laterally for long distances.
Deep-sea pelagic (abyssal hills, and abyssal ridges and rises)	Fine clay, some of windblown origin, very finely laminated, frequently with manganese nodules or crusts, interbedded with carbonate or siliceous oozes.
Dominantly chemical	
Reefs	Complex of carbonate rock types, from massive cemented reef rock of corals and/or algae to back-reef carbonate sands, lagoonal muds, and beach sands.
Carbonate platforms	Great variety of reef, lagoonal, tidal-flat, and shallow-bank deposits of oolitic and shell-fragment sand, and algal-mat and fine-grained carbonate muds.
Deoxygenated barred basins	Pyrite-rich and organic-rich muds with no bottom-dwelling marine fauna.
Restricted evaporite basins	Gypsum, anhydrite, halite, and other salts interbedded frequently with carbonates and some muds.

(From EARTH, Second Edition, by Frank Press and Raymond Siever. W. H. Freeman and Company, Copyright © 1978.)

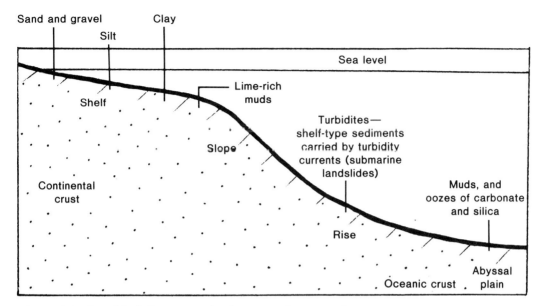

Figure 4.4. Simplified illustration of sedimentary depositional sites on a passive, Atlantic-type, continental margin.

arkose. Under slow weathering in non-tectonic settings, feldspar breaks down chemically to form clay (the main constituent of shale and mudstone).

A significant aspect of sedimentary rocks is their layering or **stratification.** Breaks between beds (bedding planes) represent pauses in the supply of sediment. The characteristics of strata (cross-bedding, graded bedding, and other sedimentary features) can indicate if the sediment making up the rock layer was deposited, for example, in a sand dune (Fig. 4.5), beach, or delta. Turbidity currents (p. 168) give graded beds (of sediments called turbidites): coarse material at the bottom of each bed (settled out first) grades upwards to fine-grained sediment at the top (overlain, in turn, by coarse material of the next bed). Rivers and wind commonly produce cross-bedding (material deposited on a slope rather than horizontally). Cross-beds and other features, such as current ripple marks, can show wind or river directions in ancient rocks and can be used to reconstruct paleoenvironments.

Features such as graded beds, mud cracks, and oscillation ripple marks (formed by waves, Fig. 13.5) can be used to indicate the top of a sedimentary layer even if the sequence has been overturned by deformation (Fig. 16.5).

As described in *Part II*, vertical and horizontal changes in sediment and fossil types within a series of sedimentary rock layers (a stratigraphic sequence) can indicate, for example, the existence and duration of erosional episodes (unconformities, page 189), or whether seas deposited sediment while advancing over, or withdrawing from, the land (transgression or regression, respectively).

Figure 4.5.   Cross-bedding preserved in lithified sand dunes, Zion National Park, Utah. (Photo courtesy U. S. Department of the Interior, National Park Service.)

Major thicknesses of sedimentary rocks commonly accumulate along continental margins in linear belts called **geosynclines.** A geosyncline, thousands of km long and hundreds of km wide, typically consists of a shallow, inner **miogeosyncline,** in which are deposited limestones, shales, and sandstones, and a deep outer part (toward the ocean) called the **eugeosyncline,** in which accumulate thicker deposits of graywacke-type sandstones (turbidites), shales, limestones, basalts and pyroclastics (see also, section on deformation, especially Figs. 16.21, 16.22 and 16.23).

Importance of Sedimentary Rocks

Sedimentary rocks and processes are important at a very practical level as they provide energy (coal, oil shale, source and reservoir rocks for oil and gas), are aquifers for underground water, include major iron, uranium and copper deposits, are used as building materials (gypsum, cement), fertilizers (potash, phosphate) and have a host of other uses.

Also, by interpreting sedimentary rocks and the fossils contained in them, we can reconstruct the geography of ancient lands and seas (paleoenvironments that help us find new resources, among other things) and establish a relative time scale (see Table 4.4).

Relative Age

Relative age is established through the application of certain basic principles that are discussed more fully in Part II. These principles include a) **Superposition** (oldest layers of rock are on the bottom of an undeformed series of strata, b) **Faunal succession** (fossil plants and animals occur in a definite evolutionary sequence through time), c) **Cross-cutting relationships** (events, such as igneous intrusions and faulting, that are younger than the rocks they cut).

Deformation and/or erosion commonly leaves isolated exposures of rocks that formerly constituted single, continuous units. To establish relative ages and the geologic history of an area, separated outcrops must be **correlated** with each other.

Correlation may involve the recognition of 1) similar rock layers, not necessarily of the same age, or 2) dissimilar layers deposited at the same time. An example of the first situation would be the prolonged, slow encroachment (transgression) of rising sea level over the land: beach sands and other sediments would be deposited over an ever-larger area; individual units might look the same but would not be deposited at the same time (a reverse distribution of sediments is produced during withdrawal [regression] of the seas). The second situation arises when there are facies changes (p. 60) or widely separated, different depositional environments.

Tracing rock units *may* be relatively straightforward because, with some obvious exceptions such as deltas, sediments are deposited in roughly horizontal layers (the principle of **original horizontality** of strata), and show **lateral continuity** (Fig. 1.1). In many areas, however, folding and faulting (section 16), as well as facies changes and erosion, make correlation extremely difficult.

TABLE 4.4 Geologic Time Scale

Era	Period	Age in Millions Of Years Ago	Biologic Events	Geologic Events
CENOZOIC	Quaternary	2		Pleistocene glaciation
CENOZOIC	Tertiary		Man Apes Elephants Horses Mammals develop rapidly	Himalayas Alps Yellowstone Volcanics
MESOZOIC	Cretaceous	65	Extinction of dinosaurs First primates Flowering plants Snakes	Rocky Mountain Orogeny Widespread oceans
MESOZOIC	Jurassic	135	Birds	Andes Present Atlantic ocean began to form.
MESOZOIC	Triassic	190	First mammals First dinosaurs Last giant amphibians Reptiles develop rapidly	Separation of Pangaea
PALEOZOIC	Permian	225	Extinction of trilobites and many other invertebrates First mammal-like reptiles	Deserts. Glaciation. Urals. Appalachian Mountains developed as Pangaea formed throughout the Upper Paleozoic.
PALEOZOIC	Pennsylvanian	280	First reptiles Flying insects Coal-forming forests	
PALEOZOIC		325		

Era	Period	Age (million years)	Life forms	Events
PALEOZOIC	Mississippian		First seed plants First reptile-like amphibians	
		350		
	Devonian		First land vertebrates (amphibians) Sharks Forests	
		400		
	Silurian		First land vegetation First air-breathing animal (scorpion-like)	
		430		
	Ordovician		Fish (first vertebrates)	Glaciation. Start of Appalachian mountain building.
		500		
	Cambrian		Many invertebrates appear	
		600		
PRECAMBRIAN	Proterozoic		Worm-like animals, algae, jellyfish, sponges	Glaciation. Primitive atmosphere deficient in oxygen. Major iron deposits. Glaciation.
		2,600		
	Archeozoic		Blue-green algae, bacteria	Oldest rock dated, 3.8 b.y. Earth begins to differentiate into core and mantle with the build-up of radiogenic heat.
		4,700	Age of the earth	

Correlation may be based on 1) **fossils** (recognition of diagnostic, so-called index fossils; or particular assemblages of several types of fossils), 2) identification of **marker horizons**— widespread, easily-recognized rock units such as layers of wind-blown volcanic ash. Where an individual rock layer is not diagnostic, a series of strata may occur in a recognizable sequence; such sequences are not unique in the geologic record, however, so care must be exercised. Other methods of correlation may involve radiometric age determinations (**p. 34**), paleomagnetic properties (**p. 214**), abundances of trace elements, seismic wave velocities, etc.

5. Metamorphic Rocks

Metamorphic rocks result when heat, pressure, and chemical reactions cause pre-existing minerals to recrystallize or combine together to produce new, different minerals without melting. Most occurrences are related to the tectonic cycle, which brings rocks into different temperature and pressure environments, and to the emplacement of intrusive igneous masses. Metamorphic temperatures range from about 150°C to 800°C (from diagenesis to the extensive development of magma).

Types of metamorphism are **regional,** which is related to mountain building (orogeny) and, hence, to plate collisions (Fig. 1.10); **contact,** where metamorphism is produced by local thermal effects around a hot igneous intrusion, giving roughly concentric zones (aureoles) of modified rocks; **burial,** at the bottom of a thick pile of sedimentary strata where metamorphism is caused by the pressure of overlying rocks and the increased temperature due to the geothermal gradient (average 3°C increase per 100 meters depth; range 1°-5°); and **dynamic** metamorphism (crushing) produced largely by mechanical effects in shear and fault zones. **Metasomatism** (page 53) is metamorphism that involves changes in the bulk chemical composition of the original material (metasomatism of limestones intruded by granitic masses has given rich deposits of iron, copper, tin and other mineral resources).

Classification of Metamorphic Rocks

Metamorphic rocks are grouped into two classes based on their gross texture; **foliated** (layered, banded, etc.) and **non-foliated.** The foliated rocks form a sequence from fine-grained to coarse-grained: slate→phyllite→schist→gneiss→migmatite (a mixed igneous and metamorphic rock representing a transition from metamorphic to igneous processes). For non-foliated rocks, texture is non-directional and less diagnostic, hence the rock is named from its composition; e.g., quartzite, marble, anthracite (note this approach parallels that for sedimentary rocks—detrital by texture, chemical by composition).

A classification of metamorphic rocks is given in Table 5.1.

Interpretation of Metamorphic Rocks

In examining metamorphic rocks one seeks to know:

a. The **original material** (this could affect our interpretation of the geologic evolution of the area—igneous versus sedimentary environments for example—and also could have very practical ramifications in terms of economically important deposits that might be expected).

Table 5.2 summarizes the sorts of rocks and new minerals that would develop during metamorphism of various starting materials.

TABLE 5.1 Metamorphic Rock Classification

One group shows compositional banding, layering or other planar surfaces. These are called foliated rocks. The other group does not show foliation. In general, although there are many exceptions, regionally metamorphosed rocks are foliated (responding to directed pressures) whereas those resulting from contact metamorphism are non-foliated (note, however, that certain minerals are equidimensional and do not develop planar orientations even in regional metamorphism).

a) **Foliated rocks** are named from their overall texture: fine-grained rocks with well-developed planar surfaces (like a deck of cards) grading to coarse-grained rocks showing alternating light and dark compositional bands. (Dynamic metamorphism produces special types of foliated rocks.)

	Name	Characteristics
Generally increasing temperature and pressure (grade of metamorphism)	Fine grained	Slate	Looks like shale but "rings" when struck. Slaty cleavage well developed.
	Phyllite	Wavy surfaces having a satiny sheen from the small mica flakes developed.
	Medium grained	Schist	Large platy minerals like mica. May be coarse grained. Amphibolite is a variety having rod-like minerals (amphiboles) rather than platy ones. Minor compositional layering may be developed.
	Coarse grained	Gneiss	Shows compositional layering with dark layers of mica and amphibole alternating with light layers or lenses of quartz and feldspar.
		Migmatite	Gneiss with bands, streaks, patches, etc. of granitic material. A mixed igneous and metamorphic rock.
Metamorphism by granulation in fault and shear zones. Generally low temperature mechanical effects (dynamic metamorphism).	Medium to coarse grained	Cataclasite	Banded or streaky rocks becoming finer grained, even glassy, with increased granulation.
	Fine grained	Mylonite	

74

TABLE 5.1 continued

b) **Non-foliated rocks** are named for their composition.

Composition	Possible original rock that was metamorphosed	Metamorphic rock
Variable	Many different types	Hornfels
Quartz	Sandstone	Quartzite (will break across quartz crystals rather than around them as in sandstone)
Calcite or dolomite	Limestone or dolostone	Marble
Large rock, quartz or carbonate fragments in a finer matrix	Conglomerate or breccia	Metaconglomerate
Carbon	Coal	Anthracite
Quartz, feldspar, pyroxene	Sandstone or granitic rock	Granulite

TABLE 5.2 Evolution of some common metamorphic rocks

Starting composition (original rock)	Foliated or non-foliated	Metamorphic rocks produced by progressive metamorphism (increasing temperature and pressure).	New minerals produced by high-grade regional metamorphism
Clay-rich (pelitic) rocks (shales, mudstones)	F	Slate→ phyllite→ schist→gneiss	Muscovite, plagioclase, potassium feldspar, garnet, staurolite, kyanite, sillimanite
	N-F	Hornfels	
Carbonates (limestone, dolostone with some clays, etc.)	F	Schistose marble; amphibolite	Plagioclase, potassium feldspar, garnet, pyroxenes
	N-F	Marble	
Basaltic igneous rocks	F	Greenschist→amphibolite	Biotite, hornblende, pyroxenes, garnet, magnetite
	N-F	Greenstone; hornfels	
Quartz-feldspar-rich rocks (sandstone, granitic igneous rocks)	F	Schist→gneiss	Potassium feldspar, sillimanite, biotite, garnet
	N-F	Quartzite; granulite	
Peridotites and other magnesium-rich rocks	F	Amphibolite	Olivine, pyroxene, amphibole
	N-F	Serpentinite/Soapstone. Rocks rich in amphibole, pyroxene or olivine.	
Coal	N-F	Anthracite	

b. The **conditions of metamorphism** (temperature and pressure, or depth and tectonic setting). During the metamorphism of clay-rich (pelitic) material, for example, there is a sequence of new minerals produced; (low temperature and pressure) chlorite→biotite→garnet→staurolite→kyanite→sillimanite (high temperature and pressure). The first appearance of the next highest mineral can be used as a rough thermometer and barometer to indicate the temperature and pressure of metamorphism. These indicator minerals define **isograds,** lines of approximately equal metamorphic temperature and pressure, that can be identified in the field and represented on a map. The isograds commonly cut across different rock types, such as gneisses and schists.

As indicated in the above sequence, minerals respond to changing temperature and pressure conditions by recrystallizing or reacting together to produce different minerals that are in equilibrium with the new conditions. A given assemblage of minerals, therefore, gradually will become different assemblages as metamorphism progresses (see also, granitization, page 53).

As temperature increases, water is driven out of hydrous minerals such as clays and micas, giving water-free minerals like kyanite, sillimanite, feldspar. The resulting absence of water in the system is a major factor in preventing high-temperature minerals from changing back to the original ones when the temperature falls. (Local changes that take place under decreasing temperature are called **retrograde** metamorphic effects.)

Metamorphic Facies

Particular mineral assemblages which have been identified as being in equilibrium and widespread in occurrence are called metamorphic facies. Some of the common facies and their diagnostic minerals are given in Table 5.3.

Conditions of metamorphism in nature can be inferred by analogy with experimental laboratory studies on the behavior of minerals under controlled high-temperature and pressure conditions. Thus certain mineral assemblages (metamorphic facies) can be taken to represent particular pressure and temperature conditions and, hence, particular locations and environments within the earth. (Similar interpretations using experimental laboratory data are made for igneous and sedimentary rocks.) Facies are shown in their typical tectonic setting in Fig. 5.1. They also are shown on a pressure-temperature diagram in Fig. 5.2. The pressure-temperature boundaries between facies are determined from experimental work on the stability fields of key minerals. The series of facies developed in particular tectonic settings (with characteristic geothermal gradients) also are indicated in Fig. 5.2.

Metamorphic Fabric

So far, we have emphasized interpretations based on changes in mineralogy. However, the orientation of platy or elongated minerals (foliation or lineation, respectively, as in schists), or of closely-spaced breaks developed perpendicular to compression (cleavage, as in slates) also can give significant information. These and similar textural features, ranging from the orientation of crystal lattices to large-scale compositional layering and folds in gneisses, constitute the deformational fabric. By studying the symmetry of fabric patterns in metamorphic rocks it is possible to determine the directions of greatest and least stress (page 174). This approach relates metamorphism to large-scale deformational processes such as folding and faulting (page 175).

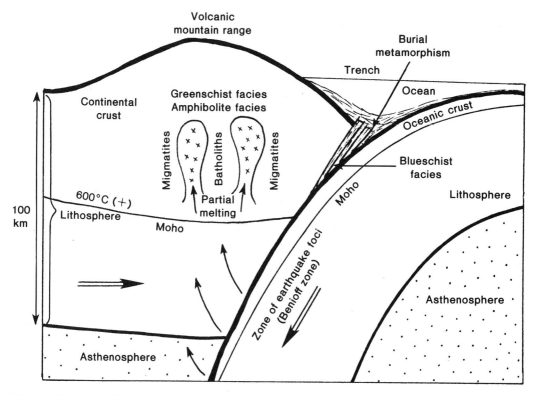

Figure 5.1. Generalized diagram of metamorphic environments during plate collision. For simplicity, associated igneous activity has been largely omitted. Metamorphic rocks occur in a paired belt: high-pressure blueschists near the trench (no granites); farther inland, medium-to low-pressure greenschists, etc., associated with granitic rocks. Around many intrusions (especially granitic masses in the upper part of the crust) there is extensive contact metamorphism. (Also shown is the Benioff zone of earthquake foci.)

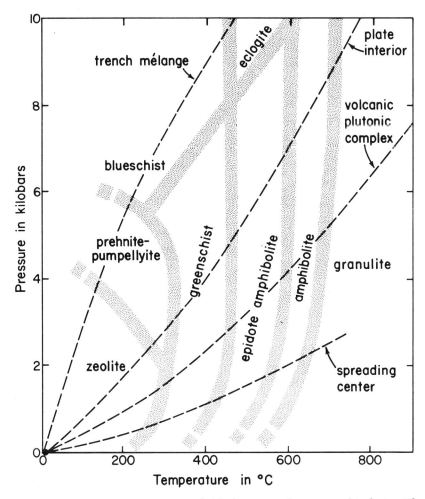

Figure 5.2. Pressure-temperature fields for regional metamorphic facies. The dashed lines indicate the facies that would be developed with increasing depth in the indicated tectonic environments. A pressure of 10 kilobars is equivalent to a depth of approximately 30-35 km (lower continental crust). Mélanges are jumbled masses of turbidite sediments and ophiolite-suite igneous rocks. (From *Petrologic Phase Equilibria* by W. G. Ernst. W. H. Freeman and Company, Copyright 1976.)

TABLE 5.3 Diagnostic minerals of selected facies produced during different types of metamorphism

Type of metamorphism	Some facies produced	Diagnostic minerals developed
Contact	Pyroxene hornfels facies	Pyroxenes but no garnet
Burial	Zeolite facies	Various zeolite minerals, e.g., laumontite
	Prehnite-pumpellyite facies	Prehnite and pumpellyite
Regional	Blueschist facies	Lawsonite, glaucophane, jadeite and quartz
	Greenschist facies	Sodic plagioclase, epidote, chlorite, (biotite)
	Amphibolite facies	Calcic plagioclase, hornblende, (garnet, staurolite, kyanite, sillimanite)
	Granulite facies	Pyroxenes with garnet, (sillimanite)

6. Mineral and Energy Resources

Mineral resources include anything that can be extracted from the earth at a profit—water, oil, sand, ores, etc., not just minerals in the strict sense. The economic geologist traditionally separates them into:

Metals—(a) precious metals—gold, silver, platinum; (b) nonferrous metals—copper, lead, zinc, tin, aluminum; (c) iron and ferroalloy metals—iron, manganese, nickel, chromium, molybdenum, tungsten, vanadium, cobalt; and (d) minor metals—arsenic, beryllium, calcium, magnesium, lithium, mercury, radium and uranium, titanium, zirconium, etc.

Nonmetals—(a) water; (b) construction materials—stone, gypsum, lime, sand and gravel, cement; (c) ceramic materials—clay, feldspar, talc, wollastonite; (d) metallurgical, chemical and refractory materials—limestone, magnesite, fluorite, sulfur, bauxite, clays, quartz, spinel; (e) industrial and manufacturing materials—asbestos, mica, talc, barite, graphite, bentonite, diatomite; (f) fertilizer materials—sulfur, potash, phosphate, nitrate, limestone; and (g) gemstones.

Fossil Fuels—coal (including peat and lignite), petroleum, natural gas. (Fossil fuels are also included in an *energy group* with nuclear, water, geothermal, tidal, solar, etc.)

Mineral resources may be separated into those used a) **to make things** (metals, plastics, cement), b) **for energy** (petroleum, natural gas, coal, uranium), and c) **to sustain life** (water, fertilizers, soil). Obviously, there is not an equally urgent need for resources in these three groups; we could make fewer "things," for example.

Metals are obtained (through smelting and refining processes) from *ore minerals* (e.g., lead from galena, Table 6.1), whereas non-metals are the actual minerals themselves and undergo comparatively little processing. Over three billion tons of metallic ores, industrial non-metals, coal, sand and gravel, etc., are mined in the United States each year (plus some nine billion tons of unwanted material called *gangue*)—approximately 85% of the tonnage is from open pits (Fig. 15.1). There are also some 600,000 producing oil wells in the U. S. Despite this level of extraction we must import vast amounts of mineral resources to meet our huge demand.

Environmental effects of mining and processing usually are immediately visible and often affect other natural resources, such as forests, wildlife, and agricultural land. Sometimes the effects are more subtle, as in groundwater pollution by industrial and mining wastes. On the positive side, reclaimed pits have provided recreational lakes.

The need for mineral resources is a derived demand—more new automobiles clearly increase the demand for metals and fuels—and we continue to use our mineral resources at an ever-increasing rate. The U. S. in the past 30 years consumed more than the entire world has since the beginning of recorded history.

Characteristics of Mineral Resources

Mineral resources differ from many other natural resources in several ways.

First, they are **non-renewable** within man's time-frame (water is a special case as it is reusable through the hydrologic cycle). The sobering prospect of depleting world supplies has led to greater care in the reuse of water, recycling and greater use of scrap (not possible for the fuels, of course), synthesis (man-made diamonds, gas from coal), and substitution (ceramics for metals). It also has spurred exploration programs for new resources as well as proposals for change in life styles—the widespread use of mass transit, for example.

Second, they are unevenly or **sporadically distributed,** creating trade and/or conflict between nations of "haves" and "have-nots"; no nation is completely self-sufficient. If the elements and fossil fuels were distributed homogeneously in the earth, they would be too diffuse for extraction at current price levels with present technology. Eight elements make up 99% of the weight of the earth's crust but these include only three of economic importance: aluminum 8%, iron 5%, magnesium 2%. All the other elements *together* constitute only 1%. Geological factors produce local concentrations of these elements (in the form of minerals). Economic concentrations range from 3-4 times the average crustal abundance for aluminum, to 100,000 times the average for mercury.

Third, most mineral resources are **underground.** They are therefore difficult to locate, and various geological, geophysical, geochemical, and remote-sensing techniques must be used in exploring for them. Sound conservation and exploration programs and certain political or economic decisions (for example, whether to fund a "crash" program to develop an alternate energy source) should be based on a full assessment of our reserves of currently-used materials and the availability of proposed alternatives. However, the hidden nature of most mineral resources creates considerable uncertainty and produces widely varying estimates and responses.

Fourth, man has virtually **no control over the location of mineral resources**; the processes of formation are not analogous to reforestation, irrigation or restocking. The political and economic impact of this simple fact can be traced through the history of conquests and control of territory endowed with valuable resources. A current example involves sovereign countries extending territorial control seaward to the 200-mile limit to include the economically significant continental shelves and slopes. Also, industrialized nations such as the U. S. are increasingly forced to rely on mineral resources from under-developed countries; with burgeoning nationalism, such arrangements obviously are not secure. (We import 75% to 100% of our needed platinum, chromium, bauxite (aluminum), manganese, tin, nickel.)

Table 6.1 summarizes the occurrences and uses of major mineral resources.

Origin of Mineral Resources

Most, but not all, *metallic ores* are associated with **igneous rocks** and originate directly or indirectly from magma which is generated by partial melting deep in or below the earth's crust. Mineralization includes segregations within the igneous body as well as late-stage in-

TABLE 6.1 Principal types of economic mineral deposit

Mineral deposit	Typical minerals	Geological occurrence	Uses	Major deposits Remarks
METALS PRESENT IN MAJOR AMOUNTS IN EARTH'S CRUST				
Iron	Hematite, Fe_2O_3 Magnetite, Fe_3O_4 Limonite, $FeO(OH)$	Sedimentary banded iron formation Contact metamorphic Magmatic segregation Sedimentary bog iron ore	Manufactured materials, construction, etc.	Mesabi, Minn.; Cornwall, Pa.; Kiruna, Sweden Resources immense; economics determines exploitation.
Aluminum	Gibbsite, $Al(OH)_3$ Diaspore, $AlO(OH)$	Bauxite: residual soils formed by deep chemical weathering	Lightweight manufactured materials	Jamaica Resources great, but expensive to smelt.
Magnesium	Dolomite, $CaMg(CO_3)_2$ Magnesite, $MgCO_3$	Dissolved in sea water Hydrothermal veins, limestones	Lightweight alloy metal, insulators, chemical raw material	Most extracted from sea water; unlimited supply.
Titanium	Ilmenite, $FeTiO_3$ Rutile, TiO_2	Magmatic segregations Placers	High-temperature alloys; paint pigment	Allard Lake, Quebec; Kerala, India Reserves large in relation to demand.
Chromium	Chromite, $(Mg, Fe)_2CrO_4$	Magmatic segregations of mafic and ultramafic rocks	Steel alloys	Bushveldt, S. Africa Extensive reserves in a number of large deposits.
Manganese	Pyrolusite, MnO_2	Chemical sedimentary deposits, residual weathering deposits, sea-floor nodules	Essential to steel making	Ukraine, U.S.S.R. World's land resources moderate, but seafloor deposits immerse.
METALS PRESENT IN MINOR AMOUNTS IN EARTH'S CRUST				
Copper	Covellite, CuS Chalcocite, Cu_2S Digenite, Cu_9S_5 Chalcopyrite, $CuFeS_2$ Bornite, Cu_5FeS_4	Porphyry copper deposits Hydrothermal veins Contact metamorphic Sedimentary deposits in shales (Kupferschiefer type)	Electrical wire and other products	Bingham Canyon, Utah: Kuperschiefer: Germany; Poland
Lead	Galena, PbS	Hydrothermal (replacement) Contact metamorphic Sedimentary deposits (Kupferschiefer type)	Storage batteries, gasoline additive (tetraethyl lead)	Mississippi Valley: Broken Hill, Australia Large resources, Many lower-grade deposits.
Zinc	Sphalerite, ZnS	Same as lead	Alloy metal	Same as lead
Nickel	Pentlandite, $(Ni, Fe)_9S_8$ Garnierite, $Ni_3Si_2O_5(OH)_4$	Magmatic segregations Residual weathering deposits	Alloy metal	Sudbury, Ontario High grade ores limited; large resources of low-grade ores; also in sea-floor Mn nodules.
Silver	Argentite, Ag_2S In solid solution in copper, lead, and zinc sulfides	Hydrothermal veins with lead, zinc, and copper	Photographic chemicals; electrical equipment	Most produced as by-product of copper, lead, and zinc recovery.

(From EARTH, Second Edition, by Frank Press and Raymond Siever. W. H. Freeman and Company, Copyright © 1978.)

TABLE 6.1   continued

Mineral deposit	Typical minerals	Geological occurrence	Uses	Major deposits Remarks
Mercury	Cinnabar, HgS	Hydrothermal veins	Electrical equipment, pharmaceuticals	*Almadén, Spain* Few high-grade deposits with limited reserves.
Platinum	Native metal	Magmatic segregations (mafic rocks) Placers	Chemical industry; electrical; alloying metal	*Bushveldt, S. Africa* Large reserves in relation to demand.
Gold	Native metal	Hydrothermal veins Placers	Coinage; dentistry, jewelry	*Witwatersrand, S. Africa* Reserves concentrated in a few larger deposits.
NONMETALS				
Salt	Halite, NaCl	Evaporite deposits Salt domes	Food;chemicals	Resources unlimited; economics determines exploitation
Phosphate rock	Apatite, $Ca_5(PO_4)_3OH$	Marine phosphatic sedimentary rock Residual concentrations of nodules	Fertilizer	*Florida* High-grade deposits limited but extensive resources of low grade deposits.
Sulfur	Native sulfur Sulfide ore minerals	Caprock of salt domes (main source) Hydrothermal and sedimentary sulfides	Fertilizer manufacture; chemical industry	*Texas: Louisiana; Sicily.* Native sulfur reserves limited but immense resources of sulfides.
Potassium	Sylvite, KCl Carnallite, $KCl \cdot MgCl_2 \cdot 6H_2O$	Evaporite deposits	Fertilizer	*Carlsbad, New Mexico* Great resources of rich deposits.
Diamond	Diamond, C	Kimberlite pipes Placers	Industrial abrasives	*Kimberly, S. Africa* Synthetic diamond now commercially available.
Gypsum	Gypsum, $CaSO_4 \cdot 2H_2O$ Anhydrite, $CaSO_4$	Evaporite deposits	Plaster	Immense resources widely distributed.
Limestone	Calcite, $CaCO_3$ Dolomite, $CaMg(CO_3)_2$	Sedimentary carbonate rocks	Building stone; Agricultural lime; cement	Widely distributed; transportation a major cost.
Clay	Kaolinite $Al_2Si_2O_5(OH)_4$ Montmorillonite* Illite*	Residual weathering deposits; sedimentary clays and shales	Ceramics: china, electrical; structural tile	Many large pure deposits; immense reserves of all grades.
Asbestos	Chrysotile, $Mg_3Si_2O_5(OH)_4$	Ultramafic rocks altered and hydrated in near-surface crustal zones	Nonflammable fibers and products	*Southeastern Quebec* Limited high grade reserves but great low grade reserves

*Formula highly variable; a hydrous aluminum silicate with other cations, such as Na, K, Ca, Mg.

trusions of volatile-rich magma (pegmatites, shown in Fig. 3.16), and material transported by hot fluids from the igneous mass into the surrounding rocks (hydrothermal deposits). A simplified picture of these processes is given in Fig. 6.1.

Many belts of mineralization occur in broadly predictable areas relative to plate boundaries, shown in a generalized way in Fig. 6.2. Deposits currently exploited include material formed up to billions of years ago even though the present plate configurations are only 200 million years old. Exploration for deposits thus requires that we can identify *old* plate boundaries in continental areas. Folded mountain ranges mark former collisions; e.g., the Appalachians. Where erosion has removed any such topographic expression, as in the Precambrian

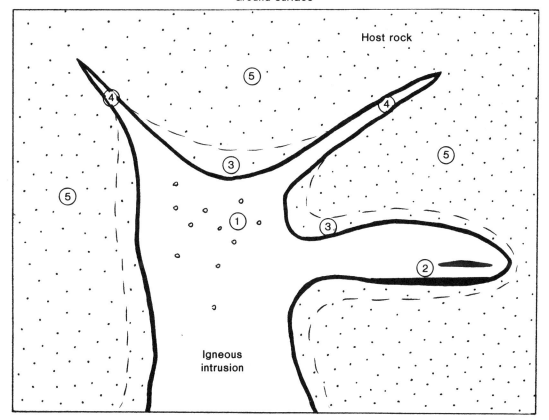

Figure 6.1. Generalized relationships among types of ore-deposits associated with igneous intrusion. (1) economically-important minerals disseminated throughout non-valuable material, (2) layers or lenses of minerals segregated by gravity or liquid immiscibility, (3) zone of contact metasomatism where the host rock is modified by high temperature fluids from the intrusion, (4) pegmatites, the volatile-rich, late-crystallizing part of an intrusion (5) hydrothermal deposits produced where hot liquids from the intrusion react with various rocks and, perhaps, groundwater.

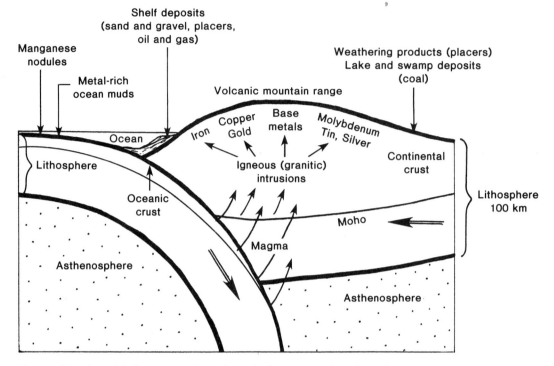

Figure 6.2. Simplified cross-section of an Andean-type plate boundary (continent-bearing plate and oceanic plate) showing the general distribution of some important mineral resources.

shields, the rock record can be used—the presence of ophiolite suites on continents is taken to indicate former oceanic crust that marks an earlier plate boundary.

The **fossil fuels,** many of the **non-metals,** and some important **metals** (iron, copper) are produced by surface **sedimentary processes,** such as weathering of rock (some laterite soils are mined for bauxite, iron, and nickel), and deposition or precipitation in swamps, lakes, or the oceans (coal, evaporites, etc.). **Placers** are sand and gravel deposits containing economic concentrations of heavy minerals—gold, diamonds, tin, etc. These minerals collect in rivers, along beaches and in similar environments where there is sufficient energy to transport lighter fragments while leaving behind the heavier ones. In this way, a concentrated deposit can be produced by the weathering and erosion of a low-grade ore body. Some deposits (e.g., uranium, copper) are concentrated by groundwater circulation.

Uranium in igneous rocks (commonly present as the mineral uraninite, or pitchblende) forms soluble compounds under weathering (oxidizing) conditions. In this state, uranium is transported by groundwater until it encounters reducing conditions and is precipitated, or enters the crystal lattice of other minerals, such as apatite in phosphate deposits. Most uranium is produced from sandstones, conglomerates and igneous veins, with organic shales providing large, low-grade reserves.

The uranium isotope U-235 is spontaneously *fissionable* (i.e., splits) producing lighter elements and releasing energy (page 34). Sustained fission is called a chain reaction and was first accomplished in 1942, an event which resulted in the atom bomb and nuclear power plants. One gram of U-235 is the energy equivalent of 2.7 metric tons of coal or 13.7 barrels of oil. Only 0.7% of natural uranium consists of U-235; the rest, mainly U-238, is not naturally fissionable. However, U-238 (and thorium-232) can be *made* fissionable through neutron bombardment (U-238 becomes fissionable plutonium-239; Th-232 becomes fissionable uranium-233). In this way, neutrons produced by fission are used to make new, fissionable material, hence the name, breeder-type reactor. There are deep concerns raised by the extremely toxic nature and very long half-life of plutonium, as well as by the prospect of nuclear weapons being made from it.

Nuclear *fusion* involves the *joining* of atomic nuclei and is the process (hydrogen joining to form helium with the release of energy) that gives solar radiation, which is seen as a major answer to many of society's energy problems. (The H-bomb fuses deuterium, an isotope of hydrogen, to form helium; controlled fusion as an energy source as yet has not been achieved.)

Oil and gas accumulations require a **source rock** (e.g., marine shale), a **reservoir rock** (permeable sandstone or limestone) into which the oil migrates from the source rock, and a **trap** to localize the deposit. The trap commonly involves beds of impermeable rock such as shale, and geologic structures (Figs. 6.3 and 6.4) that prevent further migration. Sometimes the trap is not structurally controlled, but is stratigraphic, involving the interfingering of permeable and impermeable rock layers. Figure 6.5 outlines the development of oil and gas from organic material, pointing up the significance of the depth of burial. *Oil shale* is an organic-rich lake or marine deposit that formed in a reducing environment and from which the oil did not migrate.

Coal, a major resource (Fig. 6.6), is formed from accumulated plant-remains deposited in swamps, deltas, etc. Under compaction the plants lose moisture and volatiles, changing from peat through lignite to bituminous and anthracite coal (bituminous coal has more than twice the heating value of lignite). Sulfur content is high where the coal formed in marine rather than fresh water environments.

Water, a vital mineral resource, is discussed in the sections on Rivers and Groundwater.

The Future

Without doubt we face shortages of certain mineral and energy resources, but there is considerable debate as to the causes and extent. Do the shortages reflect real geologic limitations or are they created by our present political and economic system of exploration and exploitation? In time of national emergency, should we be dependent on other countries as we are now—if so, which ones? How much should we pay for environmental concerns—non-use of available but high-sulfur fuels, delays in building nuclear power plants? The problem is a complex one.

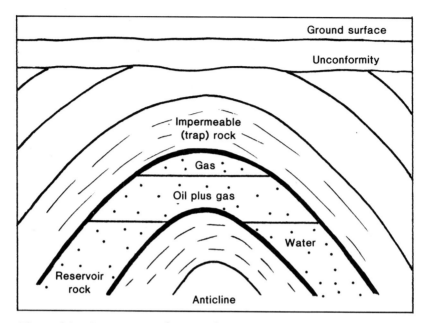

Figure 6.3. Cross-section of an anticlinal oil and gas trap. This type of trap accounts for some 80% of the oil and gas located.

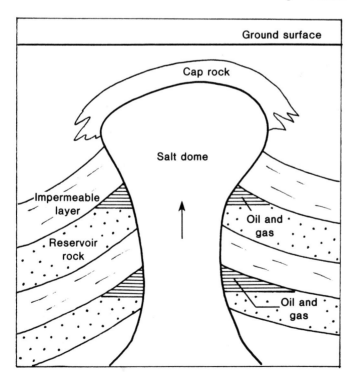

Figure 6.4. Simplified cross section of a salt dome oil and gas trap. Deep layers of rock salt are mobilized and squeezed upwards by the pressure of overlying rocks. The cap rock typically is anhydrite (calcium sulfate) which is reduced by bacteria to elemental sulfur, another valuable mineral resource. Faults above and along the flanks of the salt dome are common and provide further traps (not shown).

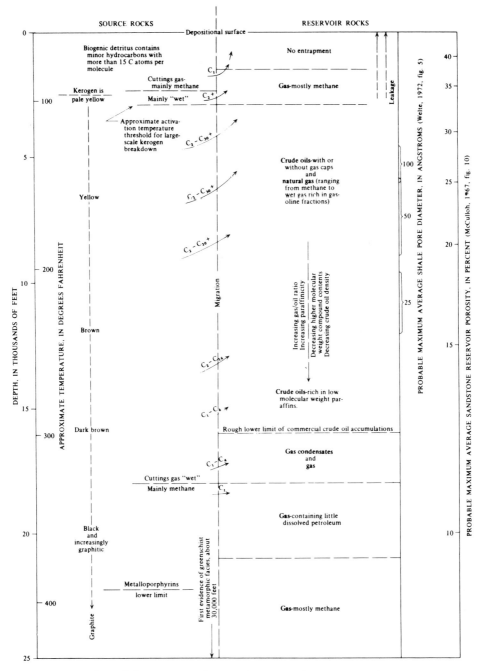

Figure 6.5. Selected diagnostic and critical steps in the genesis and migration of petroleum and natural gas from sedimentary organic matter. (From T. H. McCulloh, U. S. Geological Survey Professional Paper 820, 1973.) Kerogen (material in oil shales) consists of carbon, hydrogen, oxygen, nitrogen and sulfur; more oxygen and nitrogen than in crude oil. Paraffins are straight-chain hydrocarbons with carbon atoms connected to each other or to hydrogen by single covalent bonds (a so-called saturated series); the simplest example is methane, CH_4. Porphyrins are complex hydrocarbons derived from the pigments in blood and leaves (animal and plant remains).

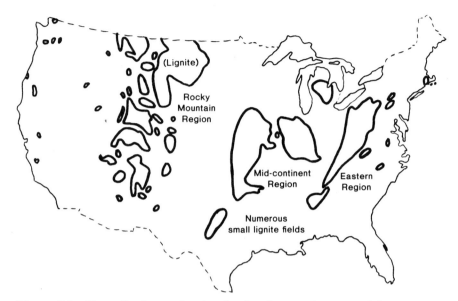

Figure 6.6. Generalized map showing the distribution of major coal deposits in the U. S. The Mid-Continent and Eastern coal areas are predominantly bituminous coal; the Rocky Mountain coals are lignite and bituminous. (Based on U. S. Geological Survey map.)

Some see a solution to mineral shortages in the fact that one cubic mile of **seawater** contains 165 million tons of dissolved solids and, for 17 critical metals, has a value of about $1,000,000. At present, seawater provides significant amounts of only a few commodities, such as salt, bromine, and magnesium. It is, of course, a source of fresh water through desalinization processes. Others point out that one cubic mile of **average rock** would give one billion tons of aluminum, 625 millions tons of iron, 12 million tons of manganese, and even 60 tons of gold. However, the energy demands and technological and environmental problems involved in extracting these resources are enormous.

Another hope lies in potential **ocean floor resources:** on and under the *continental shelf—*placers of heavy minerals (diamonds, tin, and gold), shells, sand and gravel, iron, coal, oil and gas, sulfur (exploitation of several of these deposits is already well-established); on the *continental slopes—*phosphorites; on the *continental rise—*oil and gas, sulfur; on the *abyssal plains—*manganese nodules that also contain copper, nickel, cobalt; on the *ridges—*metallic minerals.

Other solutions may include: population control, changes in our style and/or standard of living, use of lower-grade deposits, increased prices, exploration for undiscovered resources (a major task for geologists), use of different kinds of deposits (oil shale for petroleum) and energy sources (solar, ocean-temperature differences, wind, etc.), as well as stressing conservation, use of renewable resources (e.g., alcohol, wood) whenever possible, scrap recovery, synthesis and substitution, and the use of elements or minerals not now commercially exploited.

SURFACE PROCESSES AND EROSIONAL AGENTS

Landscape Arch (88 meters across). Arches National Park, Utah. Courtesy of National Park Service, U. S. Department of the Interior.

Surface Processes and Erosional Agents

We have considered the hydrologic cycle (Fig. 1.9) as part of the rock cycle, especially in the production of regolith and sedimentary rocks. In the following sections, many of the surface processes discussed provide further examples of the geologic significance of the hydrologic cycle.

Landscape is constantly changing as external processes such as weathering, mass wasting, running water, ice, wind, and ocean waves and currents attempt to reach equilibrium with internal processes such as faulting, volcanic activity and mountain-building (orogeny). Factors that control the landforms produced include relief (differences in elevation), climate, and rock types, as well as the particular internal and external processes and the rates at which they operate (Compare, for example, Figs. 1.7, 7.2, 9.13, 11.8, 16.12).

As each erosional agent is discussed in the following sections, material will be presented on five interrelated aspects: a) certain *processes* operate to give *products* that are b) *erosional* (weathering plus transportation) or c) *depositional* features. These features consist mainly of d) *bedrock* shaped in certain ways, or e) *sediment* having certain characteristics (type of bedding, size, shape, sorting of grains, etc.; for example, sediment becomes rounded and sorted if transported by water and wind but not if transported by ice or mass wasting).

Gravity is significant in all erosional processes as material (rock, regolith, water, ice) moves downslope under its influence.

Plate tectonics also affects erosional agents, for example:

a. As new basaltic crust is added at the ridges, differing rates of sea-floor spreading can cause world-wide (**eustatic**) sea level changes; such sea level changes will alter the position of shorelines (depositing transgressive or regressive sequences of sediments) and change base level for rivers. (Changing the base level of rivers determines the extent of **downcutting** or deposition and, consequently, affects the appearance of river valleys and the supply of sediment to beaches.)

b. Continents moved into different latitudes by plate motion will experience different climates, hence different weathering rates and products. Also, as wind direction and strength change with latitude, the significance of wind as an erosional agent will vary.

c. Ocean circulation patterns will be affected by the changing position of landmasses (e.g., joining of North and South America to block the equatorial current). This, in turn, could influence climate (e.g., Gulf Stream) and, perhaps, contribute to the onset of glaciation which then would affect sea level, base level for rivers, etc.

7. Weathering

As outlined on page 59, weathering involves the production of regolith (and soils) and is the response of lithosphere material to conditions in the atmosphere and hydrosphere. There are two types of weathering. One type of weathering, **physical** or **mechanical**, generally increases the surface area by breaking the rock into smaller fragments through frost-wedging, heating and cooling, etc., so that the second type, **chemical** weathering, can operate more effectively. The major agent of chemical weathering is rain, a very weak solution of carbonic acid (H_2CO_3). Fig. 7.1 shows the increase in surface area produced by breaking a cube into smaller pieces. One result of physical weathering is **exfoliation** (Fig. 7.2), the development of sheets in rock, much like the layers of an onion. In general, minerals respond to chemical weathering by dissolving (calcite) or changing (through the addition of water) to clay min-

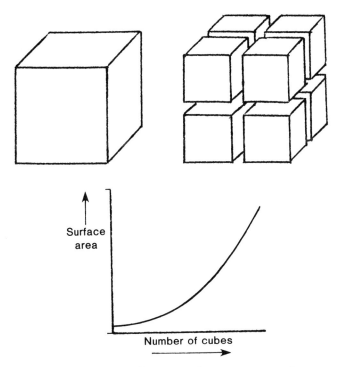

Figure 7.1. Diagram illustrating the increase in surface area when a block is broken into smaller pieces.

Figure 7.2. Exfoliation, the separation of large, curved sheets of rock by mechanical weathering and release of confining pressure. Moro Rock (granite), Sequoia National Park, California. (Photo courtesy U. S. Department of the Interior, National Park Service.)

erals, with the liberation of soluble salts, silica, and perhaps iron. Quartz is extremely stable and thus becomes relatively more abundant as weathering removes the less stable minerals (Fig. 7.3).

Table 7.1 summarizes the products of weathering common silicate minerals. The significance of weathering in the rock cycle was indicated in the discussion of sedimentary rocks (involving primarily *transported* regolith). Weathering also is essential to man's survival as productive agricultural soil is one of the results of these processes (primarily *residual* regolith plus organic matter).

The effectiveness of weathering depends on the original rock composition, the structure of that rock (joints, cracks, bedding planes), the topography (on a slope, the weathered material can be removed by gravity and rainfall; on flat ground a thicker regolith is produced), the activities of plants and animals, on time, and ultimately, on climate. (Industrial and volcanic gases accelerate weathering by producing more acid rainfall.)

Figure 7.3. Deeply weathered granite in which feldspars have decomposed to clays. Granite is overlain by Coastal Plain gravels. Columbia, South Carolina. (Photo, Jon C. Boothroyd.)

TABLE 7.1 Products of chemical weathering

Mineral		Composition	Products		
			Minerals	Others	
FERROMAGNESIANS	Olivine	$(Fe, Mg)_2 SiO_4$	Limonite Hematite Quartz (finely divided)	Some silica in solution or suspension. Carbonates of iron and magnesium in solution.	
	Pyroxene Amphibole Biotite mica	Fe, Mg, Ca, Al-silicates	Clay Calcite Limonite Hematite Quartz (finely divided)	Some silica in solution or suspension. Carbonates of calcium and magnesium in solution.	
FELDSPARS	Plagioclase	$CaAl_2Si_2O_8$ to $NaAlSi_3O_8$	Clay Quartz (finely divided) Calcite (from Ca)	Some silica in solution or suspension. Sodium and calcium carbonates in solution.	
	Orthoclase	$K\,AlSi_3O_8$	Clay Quartz (finely divided)	Some silica in solution or suspension. Potassium carbonate in solution.	
QUARTZ		SiO_2	Quartz grains	Some silica in solution or suspension.	

Soils

The result of prolonged weathering is a soil profile. Ideally, the profile consists of three roughly horizontal layers: an upper **zone of leaching** (A horizon), a middle **zone of accumulation** (B horizon), and a lower **transitional zone** (C horizon). Profiles for *pedalfer, pedocal,* and *laterite* soils are given in Figs. 7.4, 7.5, and 7.6. (In soil science a more complex scheme is used, involving ten orders plus suborders, families, etc. The names of all orders end in "sols"; aridisols, for example, are typically dry (arid) soils low in organic matter. We will use the simpler geologic groupings.)

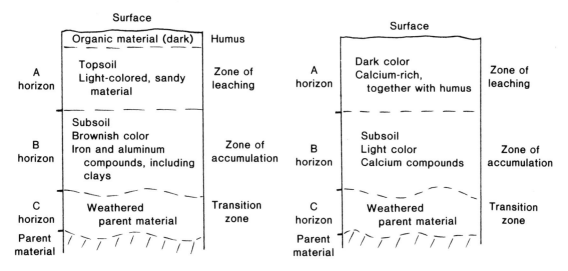

Figure 7.4. Pedalfer soil profile.   Figure 7.5. Pedocal soil profile.

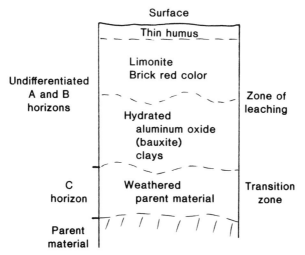

Figure 7.6. Laterite soil profile.

In the eastern U. S., soils are generally **pedalfers** (B horizon, or layer, rich in clays [alumina] and iron), while in the western U. S. the lower rainfall (less than 63 cm [25 inches] per year) does not flush the soil adequately, hence the A and B horizons are rich in calcium and the soil type is a **pedocal. Podzol** soils are varieties of pedalfers produced in temperate forests. They are characterized by a light-colored A horizon leached by waters made highly acid through decay of organic material.

In tropical climates with heavy seasonal rainfall, weathering is intense enough to break down even some of the clays by removing silica, and a particular type of soil is produced—**laterite.** Many laterites are of economic importance as sources of bauxite (aluminum), iron, and nickel.

Only *mature* soils have a well-defined profile; *immature* soils, such as those developing on regolith that was deposited by Pleistocene glaciation some 10,000 year ago, show only a C horizon, or a thin A horizon directly on top of the C horizon. The B horizon may take 100,-000 years to develop.

The **trace-element content** of soils is of great significance as it affects our health. Trace elements enter the food chain indirectly through vegetation consumed by animals, or directly from the soil into grain crops. The medical significance of trace elements is beyond the scope of this book but it is known that some elements are essential for good nutrition (cobalt, copper, molybdenum, zinc); some seem to correlate with the incidence of various diseases (iodine deficiency and goiter, fluorine deficiency and osteoporosis), while others seem to have little biological effect (tin, titanium). Many trace elements are toxic in amounts not much above the trace level. The type of bedrock, regolith and groundwater affect the distribution of these trace elements (see Section 6).

8. Mass Wasting

Mass wasting (downslope movement of rock and/or regolith) is widespread and includes generally rapid movements (commonly grouped as "landslides") and a group of generally slow movements. Mass wasting is controlled primarily by gravity (hence slope is important) and is aided by *fluids* that lubricate surfaces and weaken the material by increasing the fluid pore pressure in rocks and regolith. Fig. 8.1 illustrates some major types of mass wasting.

Slow movements include **creep**—downslope movement of soil or regolith aided by wetting and drying or freezing and thawing. **Solifluction** is a special cold-climate case of creep where the upper regolith layer thaws and becomes water-saturated while the underlying layers remain frozen and impermeable. Where the material is a mixture of regolith and ice, a **rock glacier** is formed (Fig. 8.2).

Rapid movements include material ranging from bedrock to unconsolidated debris; these materials may be water-saturated or essentially dry. Creep becomes **earthflow** with additional water; with even more water in the system a rapid **mudflow** is produced (mudflows commonly involve volcanic ash or unprotected regolith in semi-arid climates). **Debris flow** is an inclusive term. Submarine debris flows produce turbidites (Fig. 4.4).

In general, downslope movements are *flows* if they are thick (viscous) mixtures of liquid and regolith; *slides* if the material (rock or regolith) moves over the ground surface with only minor fluid involved.

Rock slides or **rock avalanches** involve rapid sliding of bedrock (and some regolith). The dip (Fig. 16.6), composition, and jointing of rock layers (strata) can be extremely important in determining their stability: beds dipping roughly parallel to a slope are more prone to sliding than are non-jointed beds perpendicular to a slope; interbedded shales are generally less stable than massive sandstones. **Slump** can involve bedrock or regolith (Fig. 8.3); in contrast to rock slides, the attitude of rock layers is not a major factor as bedrock movement is of massive blocks along curved planes of weakness.

A special category, **falls**, involves essentially free-fall of dry material (a common product is **talus**; a pile of angular rock fragments at the foot of steep slopes, dislodged by mechanical weathering).

Mass-wasting processes can be almost imperceptible, as in the case of creep, or can cause spectacular disasters with much loss of life and property damage. Man can exacerbate the natural situation by building on unstable slopes, undercutting the foot of slopes, diverting surface waters so they enter the potential slide material, stripping off protective vegetation, producing unstable piles of mining wastes, etc.

Processes Involved in Some Kinds of Mass-Wasting

Process	Definition and Characteristics	Illustration
Rockfall and debris fall	*The rapid descent of a rock mass, vertically from a cliff or by leaps down a slope.* The chief means by which taluses are maintained.	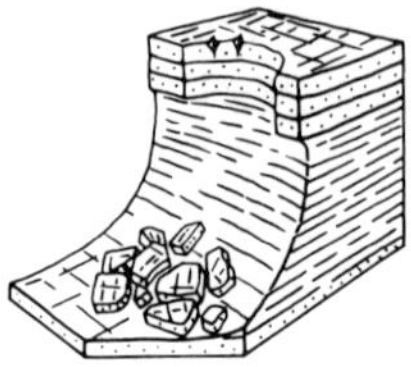
Rockslide and debris slide	*The rapid, sliding descent of a rock mass down a slope.* Commonly forms heaps and confused, irregular masses of rubble.	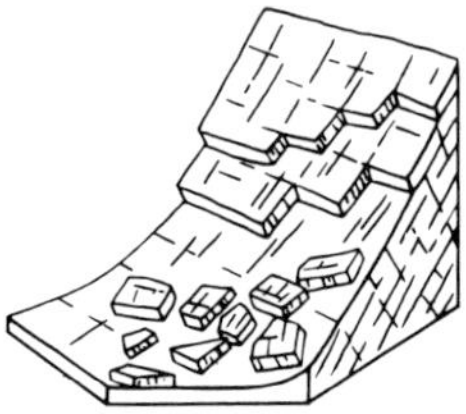
Slump	*The downward slipping of a coherent body of rock or regolith along a curved surface of rupture.* The original surface of the slumped mass, and any flat-lying planes in it, become rotated as they slide downward. The movement creates a scarp facing downslope.	
Debris flow	*The rapid downslope plastic flow of a mass of debris.* Commonly forms an apronlike or tonguelike area, with a very irregular surface. In some cases begins with slump at head, and develops concentric ridges and transverse furrows in surface of the tonguelike part.	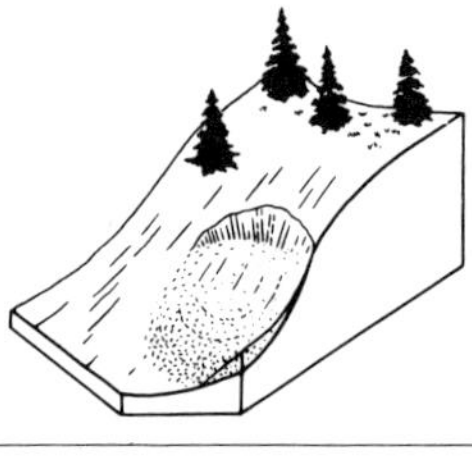
Variety: Mudflow	*A debris flow in which the consistency of the substance is that of mud;* generally contains a large proportion of fine particles, and a large amount of water.	

Figure 8.1. Some types of mass wasting. (From *Physical Geology*, Second edition, by R. F. Flint and B. J. Skinner, John Wiley and Sons, Inc., New York, Copyright 1977.)

Figure 8.2. Rock glacier, San Juan Mountains, Colorado. (Photo courtesy Gerald J. Daub.)

Figure 8.3. Small slump in highway fill, Lawrence, Massachusetts. (Photo courtesy Jon C. Boothroyd.)

Precipitation, earthquakes, or explosions can trigger massive downslope movements. Some clays liquefy under shock (quick clay) causing **subsidence** and the collapse of anything lying on them (p. 173). (Subsidence of regolith or bedrock also can be caused, for example, by excessive withdrawal of underground water, solution of limestones or other soluble material at depth, and subsurface mining.)

Mass movements of one sort or another clearly can have considerable environmental and engineering effects. Some areas of the country now require a geologist or soil scientist to certify the stability of the ground before construction is permitted.

9. Rivers

Streams are a significant part of the hydrologic cycle and are important to society as sources of energy (hydroelectric, mills), and drinking, industrial and irrigation water (reservoirs). Rivers also are valuable for trade, recreation, cooling generating plants, etc., and their floodplains provide extremely productive agricultural land.

Streams act with mass wasting to produce most of the landscape we see, as they transport the material carried downslope by various mass movements. Figure 9.1 illustrates the significance of mass wasting in the development of river valleys. (See also, Figs. 1.1, 1.7.)

Geologic Work of Rivers

Rivers act as erosional agents by downcutting their valleys, and as depositional agents by depositing materials in **alluvial fans, levees** and **floodplains,** and **deltas** (Figs. 9.2, 9.3, and 9.4, respectively). Rivers carry some 20 billion metric tons of sediment plus 4 billion metric tons of dissolved material per year to the oceans. Material carried by rivers consists of **dissolved load** (elements that went into solution during weathering and erosional processes), **suspended load** (generally fine-grained material) and **bed load** (generally coarser-grained and/or heavier material), shown in Fig. 9.5. Sediments deposited in the river channels commonly develop current ripples and cross-bedding. Sometimes, heavy particles form placer deposits of gold, tin, etc. (page 86).

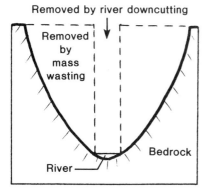

Figure 9.1. Diagram showing that river channels result from mass wasting and sheet runoff as well as from river downcutting.

Figure 9.2. Alluvial fan, Copper Canyon, Death Valley, California. (Photo courtesy Jon C. Boothroyd.)

Figure 9.3. Aerial view of a broad floodplain showing meanders, point bars, meander scars, oxbow lakes, etc. (Photo courtesy O. Don Hermes.)

Figure 9.4. Yahtse River delta, Icy Bay, Alaska. (Photo courtesy Jon C. Boothroyd.)

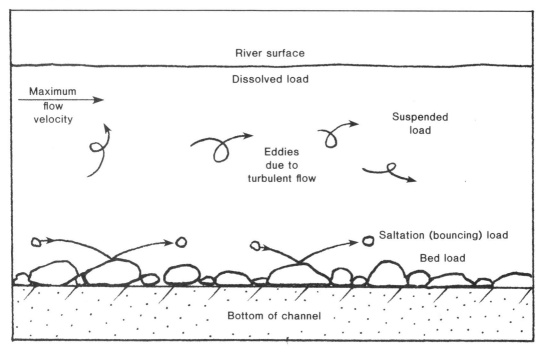

Figure 9.5. Types of load transported by a river.

The **discharge** of a river is the volume of water moving past a point in a unit amount of time. Clearly, it is influenced by, and interrelated with, the velocity of the river, the gradient (slope) of the channel, the shape of the channel, the size and shape of the bed load (frictional effects), etc. Greatest discharge and velocity occur during floods. Most erosion occurs during rising floods; deposition takes place as flood waters subside.

The ability of a river to move large (heavy) particles increases with its velocity, which in turn depends on the gradient, volume of water, shape of channel, etc. (The largest particle a stream can transport is a measure of its **competence;** its **capacity** is the total amount of sediment transported past a given point in a given time.) Braided streams result when the load is too great for the river to transport (e.g., meltwater from a glacier, shown in Fig. 9.6).

Figure 9.6. The Yana Stream, a braided stream formed by meltwater from the Malaspina Glacier (in the background), Alaska. (Photo courtesy Jon C. Boothroyd.)

Cycle of Stream Development

When considered from head to mouth, a river typically has the concave-up **longitudinal,** or **long, profile** shown in Fig. 9.7. (Cross-sections showing what the valley may look like at different places along the profile are given in Fig. 9.8).

Irregularities in the long profile will be removed with time; high parts (rapids, waterfalls) will be eroded down to the idealized profile (for example, Niagara Falls is retreating at over 1 meter per year); lower parts (lakes) will be built up through deposition of material. This is the concept of a **graded stream. Base level** is the level below which a stream can not erode; ultimate base level is sea level. If sea level fell, as during glaciation, rivers would more actively cut down to the new, lower base level in a process called **rejuvenation.** If sea level rose, deposition would occur in the drowned parts of the river valleys.

There are five possible stages of idealized stream development: (a) **Initial,** (b) **Youth,** (c) **Maturity,** (d) **Old age,** (e) **Rejuvenation.** Characteristics of each are summarized below together with features typical of the regional topography of which the stream is a part. These stages represent the classical **Davis cycle of erosion** (shown diagrammatically in Fig. 9.9). It is much oversimplified as it assumes that uplift is rapid relative to erosion and that the area remains tectonically stable through the entire cycle. Recent models stress continuous equilibrium between uplift and erosion. Despite its limitations, the Davis cycle is useful as a conceptual framework within which to view landscape development.

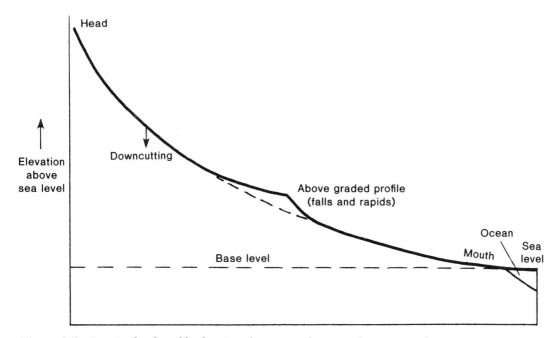

Figure 9.7. Longitudinal profile showing changes in elevation along a river from head to mouth.

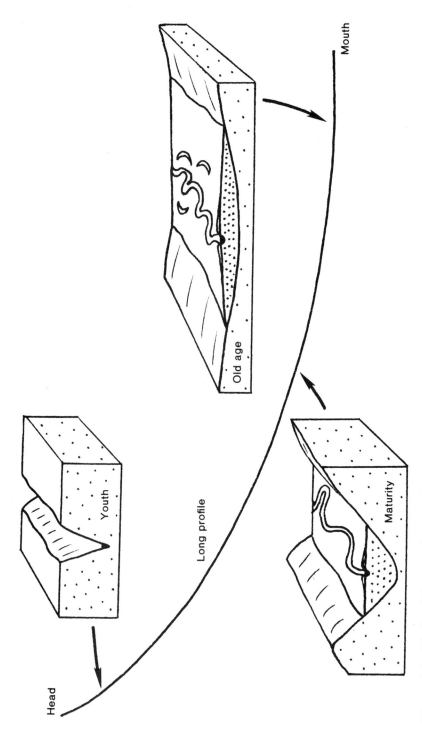

Figure 9.8. Idealized diagram showing characteristics of a river valley from head (Youth) to mouth (Old Age).

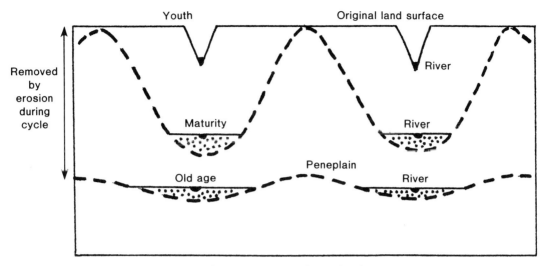

Figure 9.9. Simplified cross-section showing stages in the cycle of erosion for a given area. Note the amount of material removed from Youth to Old Age.

Initial

Major erosion is by sheet runoff on gently inclined surfaces. Streams form in low places and downcut. Lakes form in depressions and overflow along their rims. Initial stage is characterized by many small streams, waterfalls, rapids, and lakes and marshes joined by streams.

Youth

There are few tributaries; they are separated by broad divides. Valleys have V-shaped cross-profiles and are shallow or deep depending upon the height of the region above sea level. Lakes are drained when outlets are eroded down. The long profile shows many irregularities in river-bottom topography, and waterfalls (Fig. 9.10) or rapids may exist where stream channels cut across beds of particularly resistant rock; they are most typical of early youth. Downcutting and headward erosion are the dominant processes. Channels have a distinctive V-shape with steep banks. Streams are relatively straight. Steep walls provide much sediment by sheet erosion; landslides are prevalent. Steep gradients, high velocity, and bottom irregularities are typical of the youthful stage and prevent navigation of streams. Youthful streams provide hydroelectric power.

Maturity

A well-integrated drainage system is established, with many streams per unit area. Headward erosion has extended the drainage upward to the divides. Stream divides are sharp and ridge-like (in some cases, rounded) resulting in a minimum of interstream-upland. River valleys are at their greatest depth (maximum relief). Stream gradients become smooth and even; falls and rapids are removed. A profile of equilibrium has been attained by the master streams but many of the tributaries may be youthful and ungraded. Main erosive activity is

Figure 9.10. Lower Falls, Yellowstone National Park. Falls are over 100 meters high. (Photo courtesy Jon C. Boothroyd.)

sideways rather than downward—channels are widened rather than deepened. One of the main characteristics of mature streams is the **floodplain** which constitutes a considerable portion of the valley floor. Stream bends are larger and develop into **meanders** (Fig. 9.11). Full maturity is reached when the floodplain is wide enough to accommodate the meander belt. Navigation is possible because falls and rapids have been eliminated.

Old Age

Drainage is poor; streams have low gradients and meander over extensive floodplains that are wider than the meander belt (Fig. 9.3). **Oxbow lakes** (Fig. 9.12), meander scars and natural levees are more common than in the mature stage. Rivers flow on the highest part of their floodplain which sometimes prevents tributaries from joining the main stream until elevations are the same (Yazoo-type rivers). Interstream divides are low and flat. A few erosional remnants may exist as isolated hills (**monadnocks**), but the surrounding area is eroded to a **peneplain** (a relatively flat surface). Most of the surface is eroded valley floor.

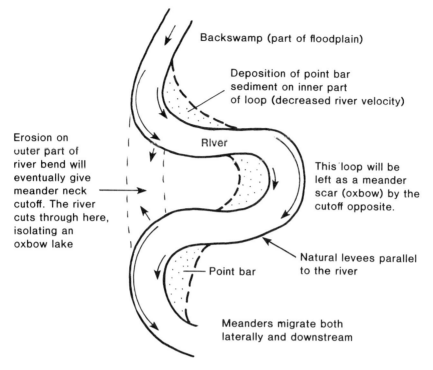

Backswamp (part of floodplain)

Deposition of point bar
sediment on inner part
of loop (decreased river velocity)

Erosion on
outer part of
river bend will
eventually give
meander neck
cutoff. The river
cuts through here,
isolating an
oxbow lake

River

This loop will be
left as a meander
scar (oxbow) by the
cutoff opposite.

Natural levees parallel
to the river

Point bar

Meanders migrate both
laterally and downstream

Figure 9.11. Diagram showing meander loops. Erosion occurs on the outside of bends, with deposition (of point bars) on the inside.

Figure 9.12. Oxbow lake in an abandoned meander loop of the Congaree River, central South Carolina. (Photo courtesy Jon C. Boothroyd.)

Rejuvenation

If land area is uplifted or base level drops, the stream will again deepen the channel by downcutting to develop terraces in the floodplain (Fig. 9.13) or even deep canyons (Fig. 9.14). Erosion begins to dissect the peneplain surface producing features typical of youth. Entrenched meanders develop because of a drop in base level; the relatively steep gradients that result are out of equilibrium with the former meander surface.

Variations in Topography

In arid climates rivers are intermittent and, during flash floods, form steep-sided box canyons (Fig. 12.2). The poorly-developed regolith and generally sparse vegetation accentuate a topography that is more angular than in moist climates. Resistant units (lavas, sandstones, limestones) form steep cliffs while less resistant shales form gentler slopes.

Limestone in moist temperate and tropical climates dissolves to form sinkholes, caverns and other solution features, commonly causing streams to disappear and flow underground (Fig. 10.13). Solution effects produce a distinctive type of topography (Karst) and a different cycle of erosion, shown in Fig. 10.11.

Figure 9.13. Terraces at two levels above the present river, produced in the Snake River floodplain by river downcutting. Grand Teton National Park, Wyoming. The face of the mountains is part of a fault scarp 65 km long. (See Fig. 16.24C.) (Photo courtesy U. S. Department of the Interior, National Park Service.)

Figure 9.14. Steep, V-shaped valley cut in Cenozoic volcanics by river rejuvenation. Rio Grande Gorge, near Taos, N. M. (Photo courtesy Reinhard K. Frohlich.)

Stream Patterns

Rivers consist of a network of tributaries, the shape of which reflects the structure and composition of the rocks over which the river flows, as shown in Fig. 9.15.

Drainage patterns and other characteristics are used to group streams into four types:

1. **Consequent** streams have a drainage pattern that depends solely on the slope of the land. The typical pattern is dendritic.
2. **Subsequent** streams develop in layers of weak rock. Because weak layers, such as shale, commonly alternate with more resistant units, like sandstone, the result is a trellis drainage pattern (where tributaries are long and parallel, following the weak layers).
3. **Antecedent** streams are older than the crustal uplift through which they cut. Such rivers flow across topographically high areas, rather than turning to flow parallel to them, because the folding or faulting that raised the area took place slowly, and *after* the river had established its course. Commonly, a mature or old-age meandering river will have a straight, V-shaped valley where it cuts through the uplift.
4. **Superposed** streams develop a pattern in a given set of strata (for example, a dendritic pattern in horizontal rock layers) but erode through those layers into rocks having a different attitude (such as tilted layers below an angular unconformity, Fig. 16.20). The stream's pattern (in this case, dendritic) thus does not reflect the attitude and composi-

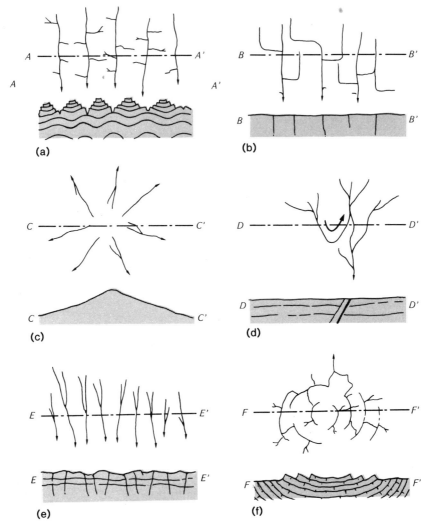

Figure 9.15. Drainage patterns and their structural and lithological controls. (Map view, above, and cross-section along the line shown, below, in each case.) (A) Trellis: controls, folds (structural) and ridges and valleys (topographic), (B) Rectangular: controls, joints or faults, (C) Radial: controls, topography (e.g., volcanic dome or cone), (D) Barbed tributaries: controls, resistant lithology (e.g., an igneous dike), (E) Parallel: controls, joints or faults, (F) Annular: controls, warped beds (structural). (From *Geology: Our Physical Environment,* by S. N. Davis, P. H. Reitan and Raymond Pestrong. McGraw-Hill, Inc., New York, Copyright 1976.)

tion of the rocks through which it now flows (a trellis pattern would be expected in the tilted layers of this example).

The area that provides water to a river channel, which is called its **drainage basin** (Fig. 9.16), is separated from adjacent drainage basins by **divides** (Fig. 11.16 shows part of the Continental Divide). Awareness of the extent of these drainage basins can be used in flood prediction and control. The drainage basin of the Mississippi River includes some 40% of the continental U. S.

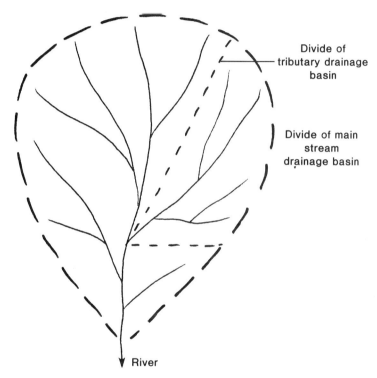

Divide of tributary drainage basin

Divide of main stream drainage basin

River

Figure 9.16. Sketch showing the drainage basin of a river system. Each segment of a river system has its own drainage basin separated from the others by divides.

Deposition

Whenever stream velocity decreases, some of a river's load may be deposited. If a mountain stream flows onto a relatively flat valley floor, an alluvial fan may be deposited (Fig. 9.2). When a river enters a large body of water (lake, ocean) sediment may be deposited in a delta (Fig. 9.4). The shape of a delta front reflects the interaction between the supply of sediment from the river and the wave and current energy acting on the delta itself (Figs. 9.17 and 9.18). Deltas commonly are many thousands of square kilometers in area. The Mississippi River delta started where Cairo, Illinois, is now and has grown some 1500 km in 70 million years.

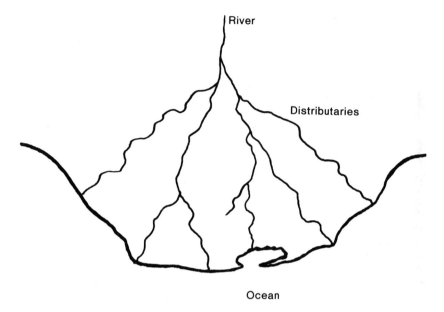

Figure 9.17. Diagram of an arcuate delta where sediment is redistributed by ocean processes, forming a smooth delta front.

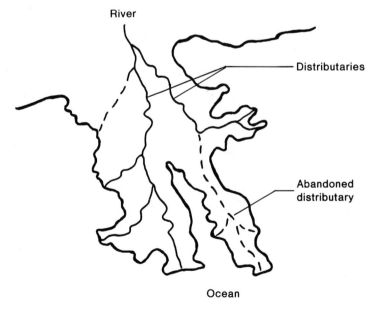

Figure 9.18. Sketch of a birdsfoot-type delta. Shape is due primarily to the sediment supplied by the distributaries (fluvial or river processes) rather than to ocean processes.

10. Groundwater

Groundwater is of considerable importance to man as a source of drinking water, and for irrigation and industrial purposes. In addition to its obvious significance in the hydrologic cycle, movement of underground water through rock and regolith dissolves and redeposits minerals as part of the rock cycle (Fig. 10.1). In this way, regolith is cemented into sedimentary rock by the deposition of silica (quartz), calcite, or hydrated iron oxide (remember the material in solution during weathering). Some economically important deposits, such as copper and uranium, are concentrated by groundwater.

Figure 10.1. Trees, originally buried by sediment and volcanic ash, replaced by silica carried in groundwater, and now exposed by erosion. Petrified Forest National Park. (Photo courtesy U. S. Department of the Interior, National Park Service.)

Groundwater exists because rocks and regolith have holes; i.e., pores and voids into which water can move from the surface. (Groundwater is present to depths of up to 10 km, but the deeper water usually is so salty it is called brine.) The ratio of pore space to total volume is the **porosity** of the material (over 15% is considered high porosity). Porosity generally decreases with depth as the pressure of overlying material compacts the rock. **Permeability** is the ability to allow fluids to flow through material. Large, connected openings, such as in well-sorted, coarse-grained sediment, give high permeability. The rate of flow depends on the permeability and the slope of the water table (hydraulic gradient). Permeable materials (sand, sandstone, some limestones) form **aquifers** (subsurface porous and permeable zones). Some materials may have naturally high porosity yet be nearly impermeable (e.g., clays and shales) if the holes within the rock or sediment are too small or are unconnected; some non-porous material, igneous rock for example, may be permeable if it is jointed or fractured. Impermeable layers are called *aquicludes*.

The Water Table

Figure 10.2 illustrates the position of the **water table** (the upper limit of the **zone of saturation**) in porous and permeable material. Note that water is distributed roughly parallel to the earth's surface. Contrary to common misconception (divining rods, etc.) only rarely (in certain limestone terrain, for example) is the water in linear "underground rivers". In areas with some topography, the water table is not horizontal and groundwater moves as indicated in Fig. 10.2. Where the water table intersects the ground surface, a spring, lake or swamp is formed.

In arid climates, the water table might be shown in Fig. 10.3. (This is the same general picture as in **recharge** when waste water is collected and used to raise a water table lowered by too much withdrawal.) In the above illustrations it has been assumed that the material be-

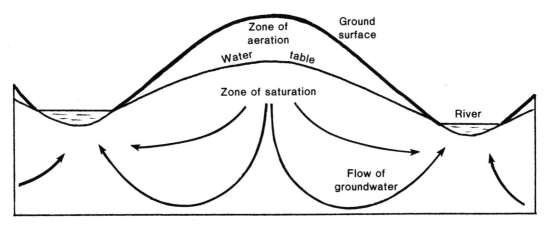

Figure 10.2. Cross-section illustrating the position of the water table and typical groundwater flow in a moist climate.

low the surface is homogeneous and can transmit water (is permeable) uniformly. If, however, there are impermeable layers then a local, or **perched, water table** may result, as shown in Fig. 10.4.

In areas near the coast, or on islands, groundwater occurs in two zones: an upper layer of fresh water that floats on a lower layer of higher-density salt water (Fig. 10.5).

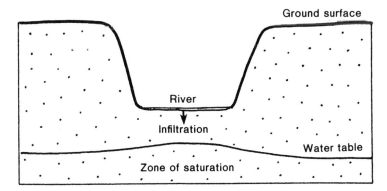

Figure 10.3. Diagram showing the position of the water table in arid regions; the intermittent stream infiltrates to the water table (contrast the situation in moist climates, Fig. 10.2).

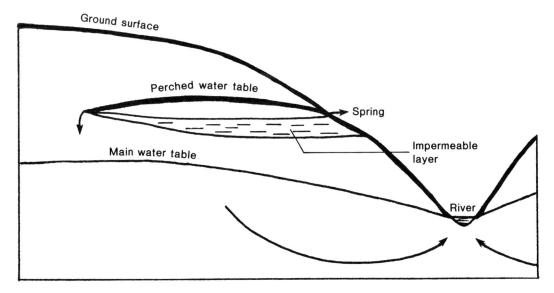

Figure 10.4. Diagram of a perched water table produced by a local impermeable layer above the main water table.

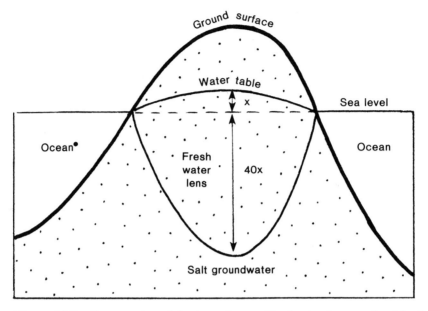

Figure 10.5. Cross-section of the groundwater distribution in coastal or island aquifers. Fresh water floats on salt water because of density differences. Note the depth of the fresh water lens below sea level is some 40 times the height of the water table above sea level.

When groundwater is withdrawn from below the water table, water adjacent to the well may be removed faster than it can be replaced by lateral flow. A **cone of depression** results, as illustrated in Fig. 10.6. If pumping continues, the cone of depression extends to the bottom of the well and it goes dry. Remember that the water table moves up and down in response to rainfall and withdrawal so wells least likely to be affected by cones of depression are those sufficiently below the lowest seasonal position of the water table. However, in the coastal environment, where fresh water lies on top of salt water, a deep well may go through the fresh water lens into the salt water below (Fig. 10.5).

If the water table is lowered significantly, compaction of the aquifer material may occur; this can permanently reduce its porosity and permeability. This compaction also can cause subsidence of the ground surface, with obvious environmental effects.

Certain wells and springs are fed by groundwater moving through rocks that have essentially no porosity (igneous and metamorphic rocks) but which are jointed and fractured giving them permeability (Fig. 10.7). These water sources tend to be more susceptible to pollution as the joints and cracks do not have the filtering effect of an ordinary aquifer.

Groundwater under sufficient pressure to rise into a well or spring is called an **artesian system** (artesian is sometimes erroneously used to describe deep or pure groundwater that is not under pressure). Requirements for an artesian well (illustrated in Fig. 10.8) are an aquifer that a) is confined by impermeable layers, b) is inclined, c) intersects the ground surface to receive precipitation, d) has sufficient difference in elevation between the water table and

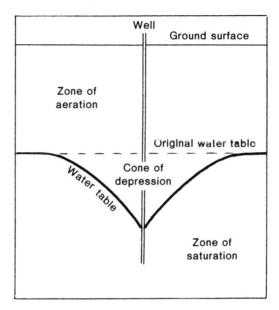

Figure 10.6. Diagram showing the cone of depression around a pumped well.

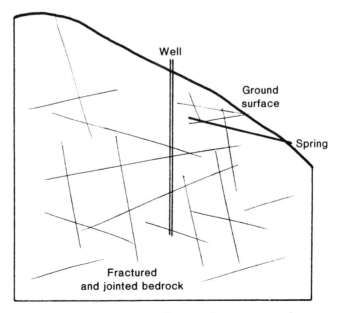

Figure 10.7. Cross-section showing that water may be obtained from nonporous rocks if they have permeability due to fractures, joints, shear zones, etc.

the well (hydrostatic head) to overcome friction in the aquifer and permit flow. Oases and other artesian springs occur where the overlying impermeable layer is breached (by a fault or erosion, for example).

The exposed recharge area of an aquifer may be tens or hundreds of km from the well or spring; at the typical rates of flow of groundwater—perhaps 1 meter per day to as little as 1 meter per year—it could take tens to thousands of years for water to make its way through an aquifer. (Slow movement of groundwater creates problems in sanitary landfills where, over a period of many years, toxic wastes are leached and distributed throughout the zone of saturation.)

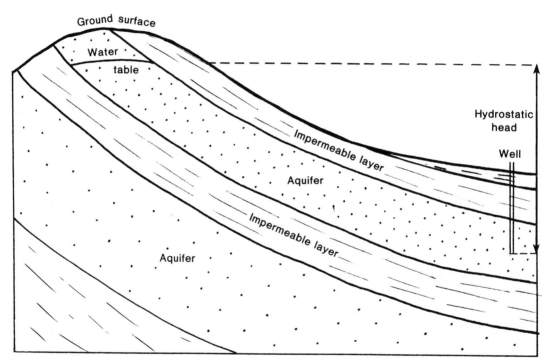

Figure 10.8. Cross-section of an artesian groundwater system.

Thermal Springs

If groundwater percolates deeply enough it can be heated either by the earth's normal geothermal gradient (some 30°C per km depth) or by coming close to bodies of recently-solidified magma in areas of young igneous activity (as most igneous activity is related to plate boundaries, we see another example of the influence of plate tectonics). In these cases, **hot springs** or the more violent variety, **geysers** (Fig. 10.9), may result. In addition to providing a spectacular display, as at Yellowstone National Park, these heated water and steam sources are important sources of **geothermal energy**. Heated underground water can trans-

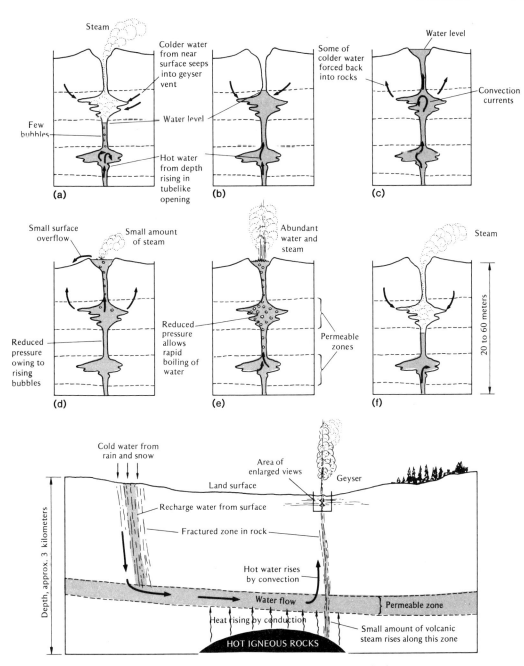

Figure 10.9. Eruption cycle of a geyser. Enlarged views are (A) end of eruption, (B) water recovery, (C) heating, (D) start of abundant bubbles in upper part of vent, (E) eruption, and (F) same as (A). (From *Geology: Our Physical Environment*, S. N. Davis, P. H. Reitan and Raymond Pestrong. McGraw-Hill, Inc., New York, Copyright 1976.)

port elements of economic importance, producing hydrothermal deposits of many important metals; deposits of silica and travertine (calcium carbonate) are common around hot springs (Fig. 10.10).

Figure 10.10. Hot spring deposits, Yellowstone National Park. (Photo courtesy Jon C. Boothroyd.)

Groundwater in Carbonate Terrain

The special case of water in carbonate (limestone; mineral, calcite) aquifers should be noted. Carried to extremes, it produces a particular topography (**Karst**) typified by sink-holes, disappearing streams, and caverns. The erosional sequence in areas underlain by lime-stone is illustrated in Fig. 10.11. Within the caverns are the well-known calcite **stalactites** (from roof) and **stalagmites** (from floor) shown in Fig. 10.12. Figure 10.13 is a topographic map of a Karst area.

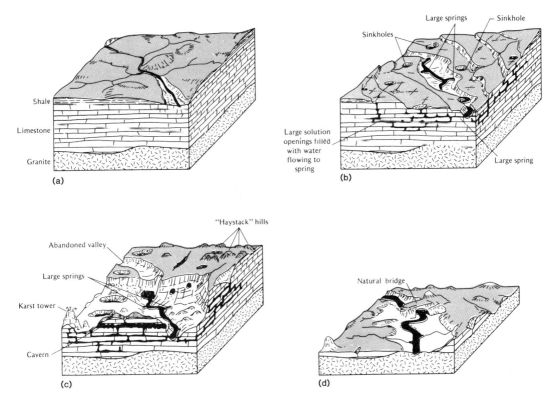

Figure 10.11. Development of Karst topography (A) Erosion removes non-permeable shale and exposes the limestone, (B) Caverns are formed by solution of water conduits beneath the water table; numerous sinkholes are formed; ground water is discharged to the surface in large springs, (C) Maximum development of caverns. Most water movement is below the surface; large rivers are the only remaining surface streams; downward cutting of rivers drains caves which were former conduits for springs; large isolated masses of limestone may form Karst towers and smaller isolated masses form "haystack" hills, (D) As limestone is removed, surface drainage is established. Some sinkholes and caverns are still preserved in the small remnants of limestone. (From *Geology: Our Physical Environment*, S. N. Davis, P. H. Reitan and Raymond Pestrong. McGraw-Hill, Inc., New York, Copyright 1976.)

Figure 10.12. Stalactites and stalagmites, Carlsbad Caverns National Park, New Mexico. (Photo courtesy U. S. Department of the Interior, National Park Service.)

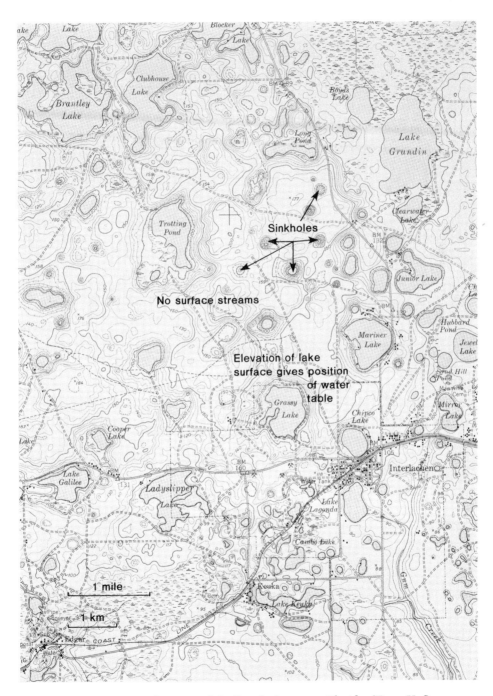

Figure 10.13. Topographic map of the Interlachen area, Florida. (From U. S. Geological Survey.)

Calcite (which is essentially insoluble in pure water) is soluble in weak acidic water, of which rainwater is an example (carbonic acid, H_2CO_3). This weak acid converts calcite ($CaCO_3$) to soluble bicarbonate, $Ca(HCO_3)_2$, and the rock is dissolved away (giving so-called **hard water**). If some carbon dioxide, CO_2, is removed from the solution by exchange with the cavern air or when drops splash onto the floor, then calcite precipitates again, leading to the formation of stalactites, etc.

(Similarly, the carbon dioxide content of sea water is important in producing **shallow-water marine carbonate rocks**—when CO_2 is removed from the water by wave agitation and warm temperatures [as in a shaken can of soda] carbonates precipitate, commonly on shell or mineral fragments. Conversely, deep [higher pressure] cold ocean waters have a high CO_2 content and carbonate shells are dissolved when they fall through the water column at those depths. Thus the very deep ocean sediments [greater than 4000 meters below sea level] do not contain carbonates; this depth is called the **calcium carbonate compensation depth** or the **CCD**.)

Groundwater as a Resource

Water is an essential natural resource. Its availability and potability are of worldwide concern and demand strict conservation, pollution-control and efficient management.

Dissolved material can be of natural origin (page 98) and/or caused by man (road salt, septic tank wastes, landfill leachate, industrial and agricultural contamination, etc.) For human consumption, dissolved material should not exceed about 500 parts per million (livestock can tolerate 2000 ppm). Commonly, taste does not indicate the presence of toxic materials and a chemical analysis is needed (a few ppm may be harmful).

An additional concern is short- and/or long-term climatic change that would reduce the precipitation available to the continents. The oceans and polar icecaps are potential sources of drinking and irrigation water; however sea water requires desalinization (which is a simple evaporation process but costly), and polar ice caps are remote (although it has been proposed that icebergs be towed to Australia, the Middle East and elsewhere).

11. Glaciation

Glaciers, which are large masses of moving ice on land, now cover some 10% of the continents. During the Pleistocene glaciation (so-called Ice Age) about 30% of the land was covered (Fig. 11.1). Glaciation is that part of the hydrologic cycle which ties up huge volumes of water that otherwise would flow back to the oceans; consequently, as glaciers expand, sea level is lowered (thus base-level for rivers is lowered, producing active downcutting of channels). The scouring action of glacial advance severely modifies existing landscapes by eroding, transporting, and depositing great volumes of sediment. The mass of ice also depresses the continents, much as a ship being loaded; when the ice melts and the glaciers retreat, the land rises (Hudson Bay has rebounded some 250 meters), and sea level rises, again affecting shorelines and river systems. (These vertical responses of the crust are examples of isostatic adjustment, described on page 195). Other glacial influences include the development of large **pluvial** lakes from increased precipitation and meltwater; development of **permafrost** (frozen ground) beyond the extent of the glacier; effects of glacial expansion and contraction on the biological record; modification of ocean circulation patterns, etc. Extensive former glaciation was first recognized by Playfair in 1802 and documented by Agassiz in 1837.

Glaciers

Glaciers result from accumulated and recrystallized snow and are moving masses of ice on land. (Fig. 11.2 shows glacier ice.) Snow, which is initially light and loosely packed (some 90% air), compacts under the weight of successive snowfalls and changes by melting, sublimation, and recrystallization into granules called **firn** or **névé**. Further compaction drives out more air and the resulting bluish glacier ice (less than 20% air) eventually (at approximately 50 meters thickness) will flow plastically downslope under the influence of gravity (Fig. 11.3); the upper zone behaves in a brittle fashion and **crevasses** form. Glaciers also move by sliding over their base (except those of the polar type which are frozen to their base; in this situation they do not erode).

The **regional snowline** crosses each glacier and is the elevation above which snow is perennial. At the present time this elevation is essentially sea level at the poles; at the equator it is approximately 4500 meters (15,000 feet) above sea level. Above the snowline the glacier receives more snow than it loses by melting and evaporation (net **accumulation**). Below the snowline the glacier has a net loss of material (**ablation**). When overall accumulation exceeds ablation, the glacier thickens and advances; when ablation exceeds accumulation the glacier thins and retreats. Glaciers can be considered as either polar or temperate types, depending on their thermal properties. Generally, however, they are discussed in terms of their shapes.

Figure 11.1. The extent of Pleistocene glaciation in the Northern Hemisphere. (By E. Antevs, courtesy, Geological Society of America.)

Figure 11.2. Glacier ice, Sheridan Glacier, Alaska. (Photo courtesy Jon C. Boothroyd.)

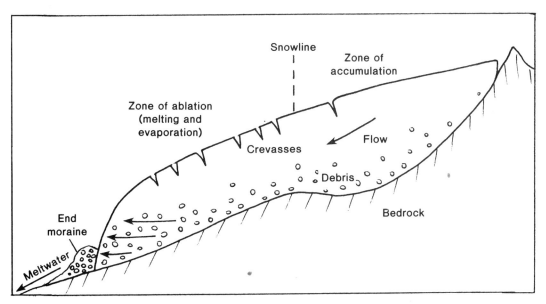

Figure 11.3. Longitudinal cross-section of a valley (or mountain or Alpine) glacier.

The two main types of glaciers are valley (or Alpine, or mountain) glaciers, and continental (or icecap, or plateau) glaciers.

1. *Valley glaciers* typically flow down pre-existing river valleys from accumulations of snow (**snowfields**) in the higher parts of the mountains (Fig. 11.4). Where several valley glaciers join, a **piedmont glacier** is formed. If the amount of snowfall increases and/or the climate get colder (but not *too* cold or the air can not hold enough moisture to be precipitated as snow), the glacier will thicken and advance down the valley; less snowfall or a warming climate will cause the glacier to thin and retreat (note the ice is still flowing down the valley even when the snout or terminus of the glacier is thinning and retreating up the valley because of excess melting). Glaciers move at an average of a few meters a year (fastest in the central part as shown in Fig. 11.5 and 11.6) but surges of several kilometers a year are observed.

As the glacier ice (and the debris within it) moves downslope it scours and deepens the valley floor. Typical **erosional** features produced by valley glaciers (recognized after the ice has melted) are shown in Fig. 11.7 (see also, Fig. 11.16) and include:

a. **Cirque** or amphitheater-like form at the head of the valley.
b. U-shaped valley cross-section shown in Fig. 11.8 (cf. V-shaped mountain river valley's cross-section).
c. Tributaries entering the main valley (deepened by glacial erosion) as waterfalls from **hanging valleys** (Fig. 11.8).

Figure 11.4. Valley glacier. Turner Glacier, Disenchantment Bay, N. E. Gulf of Alaska. Mount Cook (elevation over 4,000 meters) in the background. (Photo courtesy Jon C. Boothroyd.)

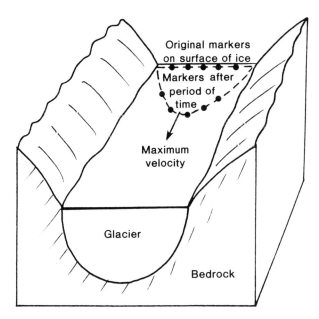

Figure 11.5. Diagram showing rate of ice movement as measured on the glacier surface. See also, Fig. 11.6.

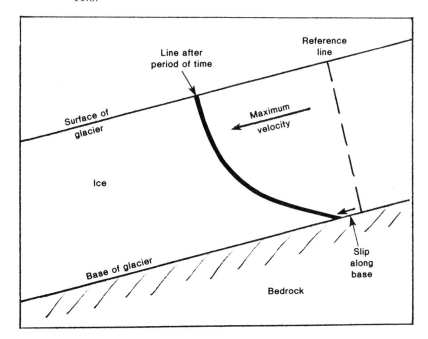

Figure 11.6. Diagram showing relative rates of ice movement in a vertical section through a glacier. See also, Fig. 11.5.

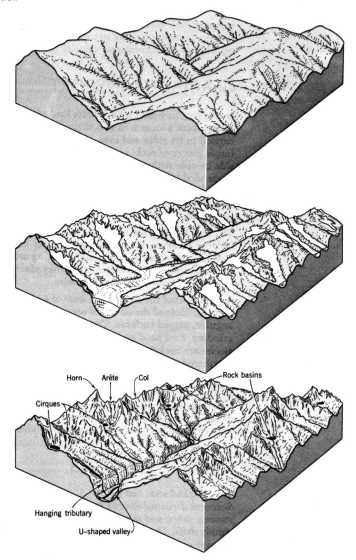

Figure 11.7. Erosion of mountains by valley glaciers. (A) Mountain region being eroded by streams. Main valley has many curves, (B) Climate grows colder, snowfields form and small cirques are excavated beneath them. Some snowfields form cirque glaciers, which merge to form a large valley glacier with tributaries. Frost wedging begins to sharpen the mountain summits, (C) Warmer climate melts glaciers and reveals their geologic work. Valleys have been deepened, widened, and straightened; tributaries are left hanging above main valley, and empty cirques, some with small lakes, now indent the highest areas. Mountain crests have been frost wedged to form knife-edge ridges (arêtes) with pyramid-shaped peaks (horns). From *Physical Geology*, 2nd Ed., by R. F. Flint and B. J. Skinner. John Wiley and Sons, Inc., New York, Copyright 1977.)

Figure 11.8. Glaciated valley showing U-shaped cross-section and hanging valleys entering as waterfalls. Yosemite National Park, California. (Photo courtesy U. S. Department of the Interior, National Park Service.)

d. Straightening of the valley and the removal of ridges or spurs along the sides.

e. Grooves, **striations,** etc., gouged into bedrock by material within the advancing ice.

f. **Roches moutonnées**—small bedrock hills shaped by plucking and abrasion (Fig. 11.9).

g. A long-profile that shows a series of steps, and perhaps small lakes (tarns), reflecting more resistant bedrock units encountered by the ice (cf. the generally smoother longitudinal profile shown by a river from head to mouth).

h. **Fiords,** valleys deepened by glacial erosion, into which the sea has now risen.

Other erosional features are seen in the mountain ranges *above* the glaciers (Fig. 11.7):

Horns—where several cirques converge from different sides into a mountain peak (examples are the Matterhorn and Mount Everest).

Arêtes—sharp, saw-tooth ridges separating the glaciated valleys.

Cols—low spots on the ridges where cirques are eroded toward each other from opposite sides of the ridge.

All material **deposited** by or derived from a glacier is called **drift;** most is unsorted and unstratified and is called **till.** Away from or against the ice, **stratified drift** may be produced by meltwater.

Depositional features are primarily moraines: till deposited as a blanket beneath the ice is called **ground moraine;** along the sides, **lateral moraine;** where two valley glaciers meet, a **medial moraine** will form from lateral moraines (Fig. 11.10). If material deposited as an **end moraine** at the terminus of the glacier (Fig. 11.12) represents the farthest advance of the ice, it is called a **terminal moraine;** if it represents a pause during the retreat of the glacier then it is a **recessional moraine.**

Beyond the terminus, meltwater sorts and transports glacial debris (often as a braided stream, Fig. 11.11) and deposits it to form an **outwash plain** of stratified drift (leaving behind the larger material).

2. *Continental glaciers* are roughly circular and up to 4000 meters thick. This mass of ice depresses the crust, which will rise by isostasy (page 195) when the ice melts (e.g., Scandina-

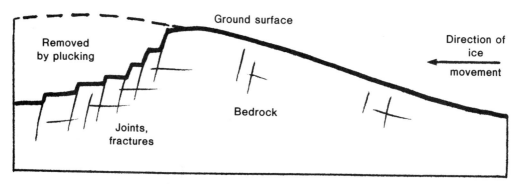

Figure 11.9. Cross-section of a roche moutonnée. The gentler bedrock slope faces the direction of ice flow. (Contrast drumlin, Fig. 11.13.)

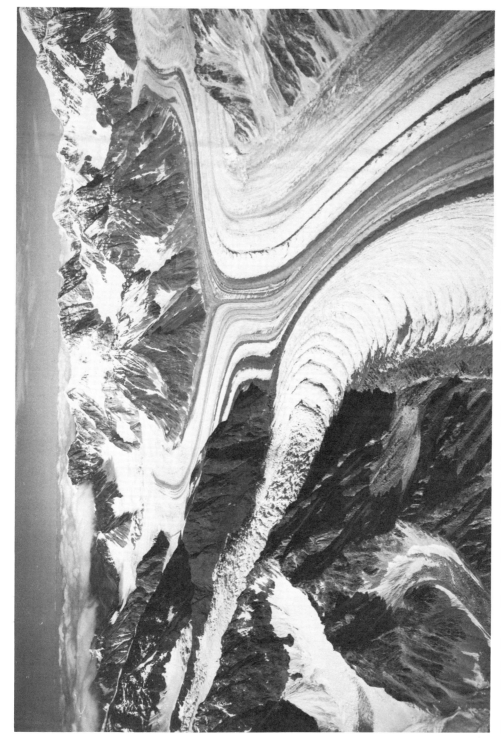

Figure 11.10. Yentna Glacier, Mount McKinley National Monument, Alaska. Note the bands of debris that produce lateral and medial moraines. (Photo courtesy U. S. Department of the Interior, National Park Service.)

Figure 11.11. Braided stream. Scott Stream, N. E. Gulf of Alaska. (Photo courtesy Jon C. Boothroyd.)

via, which is rebounding at about 1 cm/year). The Antarctic glacier contains some 90% of the earth's ice; most of the remaining 10% is in the Greenland ice sheet.

Erosional features created by continental glaciers are not so obvious as those of valley glaciers because almost the entire ground surface has been over-ridden by ice (isolated rock masses projecting above the ice are called **nunataks**), and the deglaciated landscape is mantled with drift. However, regolith and bedrock *are* eroded (to be deposited as moraines, outwash, etc.). Roches moutonnées, striations and grooves, and boulder trains of unusual rock types (**erratics**) carried from their parent outcrop can be used to trace the direction of ice movement.

Depositional features (Fig. 11.12) include moraines (ground and end, but, unlike valley glaciers, not lateral or medial as there is rarely a source of debris above the ice). Where clay-rich ground moraines are shaped by moving ice, **drumlins** are formed (Fig. 11.13). Where sediment is carried by meltwater streams, stratified drift can form against or under the ice: *ice-contact deposits* of water-transported debris are **kames**, **kame deltas** (or delta kames) or,

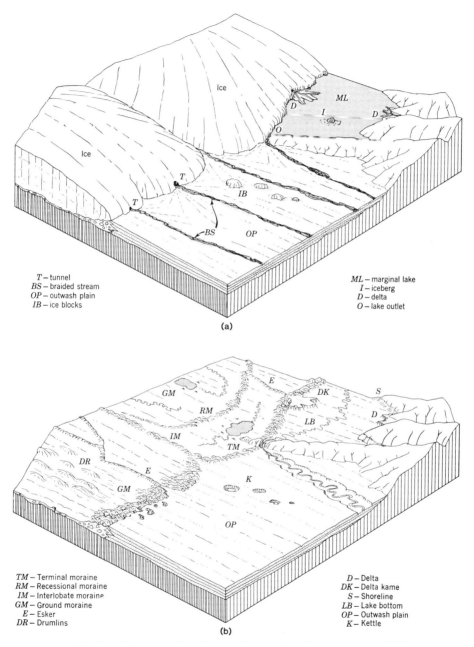

T – tunnel
BS – braided stream
OP – outwash plain
IB – ice blocks

ML – marginal lake
I – iceberg
D – delta
O – lake outlet

(a)

TM – Terminal moraine
RM – Recessional moraine
IM – Interlobate moraine
GM – Ground moraine
E – Esker
DR – Drumlins

D – Delta
DK – Delta kame
S – Shoreline
LB – Lake bottom
OP – Outwash plain
K – Kettle

(b)

Figure 11.12. Landforms produced near the margin of an ice sheet. (A) ice margin in almost stagnant condition, (B) Ice entirely gone, revealing subglacial forms. (From *Environmental Geoscience: Interaction between Natural Systems and Man,* Arthur N. Strahler and Alan H. Strahler, Hamilton Publishing Company, Santa Barbara, California. John Wiley and Sons, New York, Copyright, 1973.)

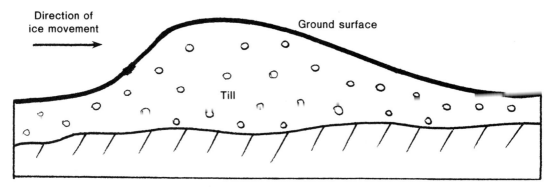

Figure 11.13. Cross-section of a drumlin. The steeper slope faces the direction of ice flow (contrast roche moutonnée, Fig. 11.9). Drumlins may be up to 50 m high and 1 km long.

if more continuous, **kame terraces**; the sinuous subglacial deposits (Fig. 11.14) are **eskers**. If ice blocks become isolated in, and covered by, outwash material, their subsequent melting will leave depressions called **kettles** (Fig. 11.15).

Sediments in glacial lakes commonly are **varved**—each varve is a pair of layers about 1 cm thick, representing an annual deposit; i.e., silt in summer, and clay deposited from suspension in the winter when the lake surface is frozen. Varves can be counted, like tree rings, to give the age of the lake, and their included pollen may indicate changes in vegetation and climate.

Summary of Glacial Features

Erosional features:
    Valley and Continental glaciation:
        Striations, roches moutonnées
    Valley glaciation (Fig. 11.16):
        Cirques, horns, arêtes, cols, U-shaped
        valleys, truncated spurs, hanging valleys, fiords

Depositional features:
    Valley and Continental glaciation:
        Till, terminal moraine, recessional moraine, ground moraine, boulder trains (erratics), stratified drift, kettles, kames
    Valley glaciation:
        Lateral moraine, medial moraine
    Continental glaciation:
        Drumlins, eskers

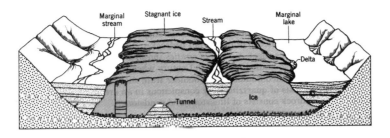

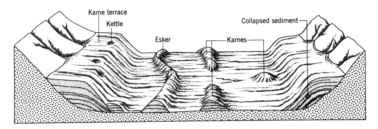

Figure 11.14. Origin of bodies of ice-contact stratified drift. (A) Nearly motionless melting ice furnishes temporary retaining walls for bodies of sediment built chiefly by streams of melt-water. (B) As ice melts, bodies of sediment slump, creating characteristic knolls, ridges, terraces, and closed depressions. (From *Physical Geology*. Second Edition, by R. F. Flint and B. J. Skinner, John Wiley and Sons, Inc., New York, Copyright 1977.)

Figure 11.15. Two kettles (dark areas) exposed by erosion of the cliffs. Block Island, Rhode Island. (Photo courtesy O. Don Hermes.)

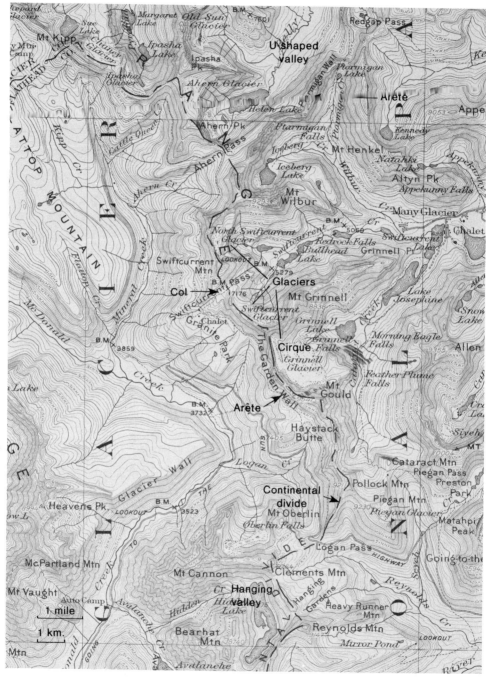

Figure 11.16. Topographic map of Chief Mountain area, Montana. (From U. S. Geological Survey.)

Ice Age

The Pleistocene epoch (Ice Age) began some 2 million years ago and consists of four glacial stages—Nebraskan (oldest), Kansan, Illinoian, and Wisconsin (youngest)—separated by longer interglacial stages when the climate was as warm as, or warmer than, the present. (The Antarctic ice sheet has existed for some 20 million years.)

Various age-dating methods (radioactive decay, magnetic polarity, fossils) are used to establish the duration of these stages. Recent data indicate that each of the three older glacial stages lasted about 150,000 years; the Wisconsin Stage lasted some 75,000 years and ended approximately 10,000 years ago (sea level at that time was over 100 meters below its present level). It is not unreasonable to consider our present climate as another interglacial (with the implication of a future ice advance). If all glaciers were to melt, sea level would rise some 60 meters, submerging many large population centers. Because each successive ice advance overrode much of the land affected by earlier advances, the glacial topography we see in the U. S. is due primarily to the Wisconsin Stage.

Older Glaciations

It is important to realize that there were glacial periods long before those of the Late Cenozoic mentioned above. This is significant as it means glaciation cannot be due simply to the earth's cooling or to any other cause that involves a unique event.

There is evidence for extensive glaciation some 275 million years ago in South America, South Africa, Australia, and India (used as evidence of continental drift as part of the plate tectonics hypothesis, see Fig. 18.5). Evidence of glaciation also is recognized in various parts of the world in rocks approximately 2.3 billion, one billion, 800 million, 600 million and 450 million years old (till lithified to tillite).

Causes of Glaciation

The fundamental cause of glaciation is not known, although certain relationships can be established; for example, temperatures seem to decrease as the strength of the earth's magnetic field decreases; moreover, glacial stages appear to coincide with magnetic reversals (p. 208) but the causal relationship is not clear. Perhaps both effects are caused by a third process. One prerequisite for glaciation seems to be continents occupying higher (colder) latitudes; the movement of landmasses into higher latitudes by plate tectonics is thus a possible initiating mechanism, although it does not account for *repeated* glaciations within one glacial age. Other causes may include:

Astronomical—variation in output of solar radiation; changes in the earth's axis and orbit (as studied in detail by M. Milankovitch).

Atmospheric—volcanic dust and gases shielding the earth from solar heat.

Oceanic—North and South America joined just before the Pleistocene thus the equatorial warm-water currents were forced northward to provide the moisture needed for precipitation of snow. The cutting off of continuous equatorial circulation also is considered as a cause of cooling.

The Ewing and Donn theory explains the cyclicity of glaciation through the availability of essential moisture (glacial advance) when the Arctic ocean is unfrozen, and the absence of moisture (glacial retreat) when the ocean is frozen. This pattern is influenced by the availability of warmer Atlantic waters: during glacial retreat, rising sea level would enter the Arctic ocean basin and provide moisture for another glacial advance; during glacial advance, sea level would become lower and the warmer Atlantic waters could not reach the Arctic basin, which would then freeze and cut off the moisture supply, thus causing retreat of the glaciers again.

12. Wind

Wind is an important erosional agent wherever dry, relatively fine-grained, unconsolidated material is exposed, such as in glacial outwash, in unprotected soil, along shorelines or in very arid regions, including the rain-shadows of high mountains. (In somewhat less arid conditions, intermittent rainfall also is a major erosional agent.) Dry regions cover about 25% of the land area; most deserts are hot but some are cold. Desert surfaces may consist of bedrock, coarse gravel, or sand.

The intensities of certain geologic processes are different in arid climates:

1. Mechanical weathering is more significant than chemical weathering, thus soils (and clays) are not well developed. This poorly developed soil profile and consequent lack of vegetation give the topography a more angular appearance (**mesas**, **buttes**, etc.). (A familiar feature produced by differential weathering and erosion of rock layers in desert areas is shown in Fig. 12.1.) Limestones form resistant cliffs whereas in moist climates they dissolve to form the solution caves, valleys, etc., of Karst topography (Fig. 10.11).

Figure 12.1. Resistant unit on top protects weaker units beneath, giving a perched rock. Santa Fe, New Mexico. (Photo courtesy Reinhard K. Frohlich.)

2. The water table commonly is deep, thus rivers lose water by infiltration. Only the largest rivers (e.g., the Nile or Colorado, which do not originate in arid areas) actually flow *through* deserts.

3. Sporadic rainfall in areas covered with loose regolith develops erosional features such as steep-sided valleys (**wadis, arroyos,** etc., Fig. 12.2) and deposits material in alluvial fans (Fig. 9.2). Debris flows are common (Fig. 8.1).

4. **Playa lakes** develop intermittently in desert basins (playas). Evaporation gives deposits of borax, sodium carbonate and other salts.

5. **Pediments** are cut into bedrock surfaces (Fig. 12.3).

6. **Wind** is a more potent erosional agent, primarily by **deflation** (the removal of sand-sized and smaller particles, leaving behind the coarser material), and **abrasion** ("sand blasting") which produces **ventifacts** (faceted boulders). The general pattern of surface winds is shown in Fig. 12.4.

The smaller products of erosion, typically particles less than 1/16 mm in diameter, are transported as dust storms and are carried into the upper atmosphere. Sand storms, as the name implies, transport mainly sand-sized (and smaller) particles; these storms are active nearer the surface of the earth where wind commonly moves the particles by **saltation** (jumping or bouncing motion). Particles much larger than sand can not be transported by wind.

Figure 12.2. Box canyon (arroyo) produced in 12 years by erosion of unprotected regolith (material shown in Fig. 7.3). Columbia, South Carolina. (Photo courtesy Jon C. Boothroyd.)

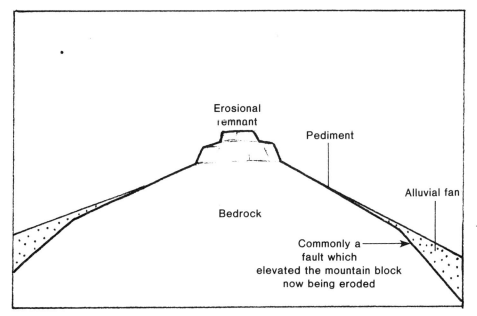

Figure 12.3. Cross-section of a pediment, an erosional surface cut into bedrock and partially covered by regolith. Isolated erosional remnants above the pediment are called inselbergs.

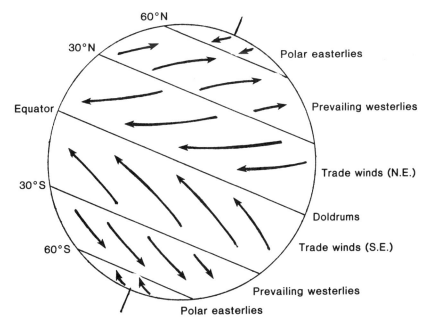

Figure 12.4. Simplified diagram showing overall wind circulation patterns.

Silt-sized materials are deposited as blankets of **loess,** whereas larger particles form **dunes** (Fig. 12.5) which migrate downwind. The type of dune depends on the amount of sand available, strength and direction of wind, type of ground surface, vegetation, etc., as shown in Fig. 12.6.

Figure 12.5. Barchan dune showing characteristic down-wind horns. N. E. of Salton Sea, California. (Photo courtesy Jon C. Boothroyd.)

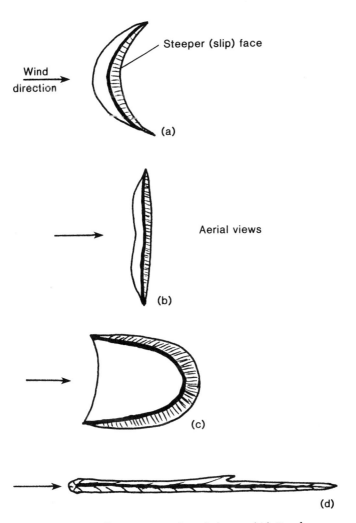

Wind direction

Steeper (slip) face

(a)

Aerial views

(b)

(c)

(d)

Figure 12.6. Different types of sand dunes. (A) Barchan. Crescent-shaped, 30 m high to 300 meters long. Requires rather constant wind direction and limited sand supply. Moves over flat, firm surfaces, (B) Transverse. Produced instead of barchans where abundant sand is available and wind is stronger. Common as irregular ridges along beaches, (C) Parabolic, also called blowout or U-shaped dune. Variation of transverse dune where some parts are stabilized by vegetation and other parts are not. Common along beaches, (D) Longitudinal. Up to 100 meters high and 100 km long. Also called seifs where shaped by cross-winds. Small amount of sand available, high winds.

13. Coastlines

The **shore** is defined as the area between low tide and high tide; the **coast** extends landward from the shore. **Beaches** are deposits of sediment moved by waves; they have been called "rivers of sand" but also include other sediment sizes.

Coastlines are influenced by many factors, especially wave (and wind) energy. **Waves** are produced by winds and radiate out from storm centers at sea. The height of the wind-generated waves depends on the wind speed, its direction and duration, and the length of open water over which the wind blows, called the *fetch*. More rarely, a different type of wave (called **tsunami**) is produced by submarine earthquakes and volcanic explosions. Tsunami can travel over 800 km/hr but are barely noticeable in the open ocean as they have wave heights of less than one meter and wavelengths up to 200 km; in shallow water they may reach heights of over 50 meters, causing much destruction.

Also affecting the development of shorelines are **tides**, which are the result of gravitational attraction between earth and moon (and, to a lesser extent, the sun) which pulls the hydrosphere closer to the moon (Fig. 13.1). The specific behavior of tides is influenced by the shape of a particular ocean basin and the coastal morphology (some tidal ranges can be harnessed as an energy source). Other influences on shorelines include the nature of the sediment and bedrock being acted on, changes in sea level, etc. World-wide changes in sea level are called **eustatic** changes. Evidence for relative change in sea level is shown in Figure 13.2.

Waves

Waves are movements of the water surface; they have various properties such as **wave height** (twice the amplitude) and **wavelength**, as shown in Fig. 13.3. In deep water, wave speed is expressed as a function of wavelength; in shallow water, as a function of the water depth. Wavelength and wave speed decrease in shallow water; wave height decreases slightly then increases; period remains constant. **Period** is the time it takes for the next crest to arrive at the same spot as the previous crest (period $= \dfrac{\text{wavelength}}{\text{velocity}}$). The period is relatively easy to measure and by a simple conversion (velocity in mph is approximately 3.5 times the period measured in seconds) one can estimate the velocity of the wave. Wave energy is proportional to the square of the wave height. When wave height approaches 1/7 the wavelength, the wave collapses; in shallow water, breakers occur when the wave height is about 0.8 times the water depth.

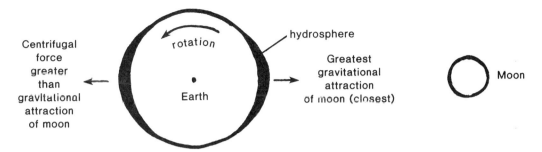

Figure 13.1. Tides result from gravitational attraction and centrifugal force. The moon (and to a lesser extent the sun) attracts the earth's hydrosphere creating a bulge facing the moon. On the earth's opposite side, centrifugal force of the earth-moon system creates another bulge where gravitational attraction is least. As the earth rotates through the bulges, there are two tides per day. The highest tides are when the moon, sun and earth are in line (full or new moon).

Figure 13.2. Changes in relative sea level in the Mediterranean recorded by water marks on the pillars. (Photo, Eugene J. Tynan.)

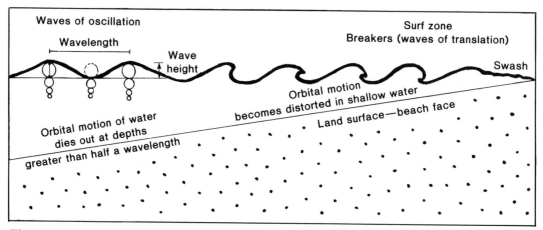

Figure 13.3. Orbital motion of water particles in water depths less than half a wavelength. In shallower water the orbital paths are deformed by frictional bottom-effects, resulting in breakers.

Water particles near the surface move in circular orbits driven by wind blowing over the water surface (Fig. 13.3), giving **waves of oscillation** (a floating object bobs up and down; it is not carried forward unless moved by the wind. Various turbines have been designed to harness this motion and convert it to electrical energy). These circular orbits diminish in radius with depth and die out at a water depth of about half the wavelength (thus material on the ocean floor in water depths greater than half the wavelength is largely unaffected by even the most violent storms).

Nearer the shore, the water depth decreases; when water depth is less than half the wavelength the water no longer can move in circular paths because of frictional drag on the bottom. The orbital paths, therefore, become more deformed as the water depth decreases and this results (when wave height is approximately 0.8 times the water depth) in the generation of **breakers**, or **waves of translation** (where there is lateral movement of the water and of any suspended object), shown in Fig. 13.3. The type of breaker (plunging, spilling, etc.) depends on the slope of the **beach face** as well as on the wave steepness (ratio of wave height to wavelength).

A breaking wave rushes up onto the beach in a turbulent mass of water called the **swash**; the water returns as **backwash**. Wave energy causes erosion, transportation, and deposition of sediment.

When onshore winds combine with the normal wave movement, there is a tendency to pile up water along the beach (*surge*). This excess buildup may be released as an off-shore flow in the form of a **rip current** (Fig. 13.4).

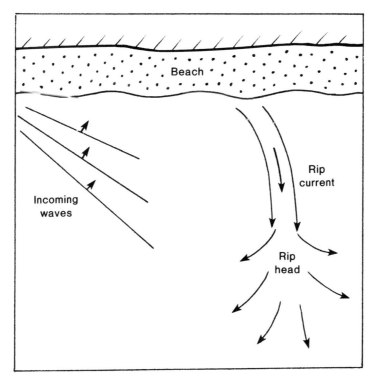

Figure 13.4. Sketch of a rip current. The current can be identified as an area where waves are barely noticeable.

Sediment Transport

The source of most beach sand is material carried to the ocean by rivers. As more reservoirs and dams are constructed, sediment is trapped upstream and the beaches are starved. Without a beach face against which to dissipate energy, storm waves can cause severe damage to coastal structures. High-energy storm waves (most common in winter) can erode the beach (Fig. 13.5) and backwash transports the removed material seaward to be stored in offshore bars. The gentler non-storm waves (most common in summer) transport the material shoreward to build a **summer berm** in front of the **winter berm**, thus extending the beach again (the berm is the nearly horizontal part of a beach formed by wave deposition).

The frictional bottom-effect that produces waves of translation also causes **refraction** (bending) of waves as they approach the shore (Fig. 13.6). The angle at which the wave approaches the shoreline gets progressively less as the end of the wave in shallow water (S in Fig. 13.6) is slowed by the bottom effect while the offshore end continues at its former velocity. The result is that waves breaking on the beach face are more nearly parallel to the shoreline than are waves offshore. Even so, waves almost always strike at some angle to the shore and move sediment in two ways: 1) across the beach face (Fig. 13.6); 2) as **longshore cur-**

Figure 13.5. Beach showing erosion of berm in the background (dark area is the break between the beach face and the berm being eroded). Plum Island, Massachusetts. (Photo courtesy Jon C. Boothroyd.)

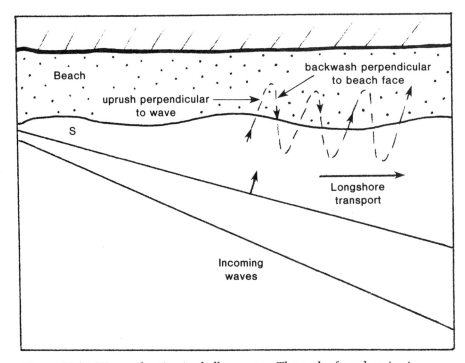

Figure 13.6. Wave refraction in shallow water. The path of sand grains is shown by the dotted line. (S indicates shallow water.)

rents which transport large volumes of sediment in the **surf zone** (Fig. 13.7). This *longshore drift* of sediment is sometimes interrupted by natural or man-made obstacles (Fig. 13.8). (At intervals along the continental shelf, submarine canyons may intercept the longshore drift and carry sediment out to the deeper ocean.) Continued movement of sand along the beach (Fig. 13.9) can cause bays and harbors to be partially or completely closed-off by an accumulation of sand in the form of **bars** or **spits** at their mouths (sometimes requiring expensive dredging operations to keep harbors open). Wave refraction and sediment transport around the end of a spit give it a characteristic curved shape. A **tombolo** is an island tied to the mainland by a sandspit (Fig. 13.16); if an island protects the shore from incoming wave energy, longshore drift is curtailed and sediment builds up behind the island (man-made breakwaters can act in the same way and cause build-up of sediment in harbors, etc.)

Spits may develop into larger accumulations of sand, typically including beaches and vegetated dunes, that are called **barrier spits** or **barrier islands**. Barrier islands form when large spits are separated into segments by tidal inlets (Fig. 13.10); they also form when a rise in sea level leaves former beach ridges and sand dunes separated from land by a lagoon.

Figure 13.7. Aerial view of the surf zone (white band). The gray area on the right is covered with sand dunes. Plum Island, Massachusetts. (Photo courtesy Jon C. Boothroyd.)

Figure 13.8. Aerial view showing how longshore drift of sediment can be interrupted by man-made groins. Plum Island, Massachusetts. (Photo courtesy Jon C. Boothroyd.)

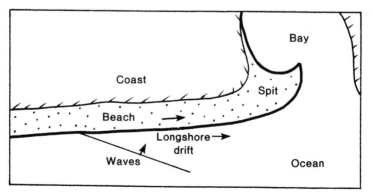

Figure 13.9. Diagram illustrating the development of a spit by longshore drift.

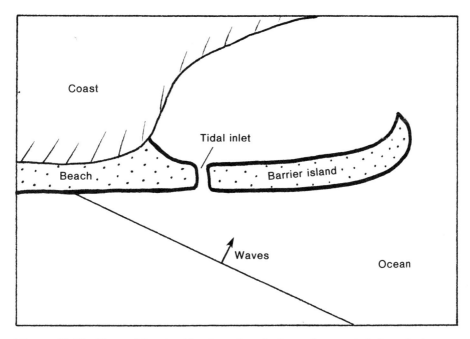

Figure 13.10.   Type of barrier island produced when a large spit is breached by a tidal inlet.

Shore Profile and Coastal Features

Waves tend to straighten coastlines (Fig. 13.11) by concentrating wave energy on headlands by refraction, thus eroding them while depositing material in the lower-energy bay areas (Fig. 13.12). Figure 13.13 is a cross-section of a typical shore profile produced by coastal erosion and depositional processes.

Erosional features of coasts include *cliffs, wave-cut benches, stacks* (Fig. 13.14), *arches* (Fig. 13.15), *sea caves*. **Depositional features** include *beaches, spits, barrier islands, tombolos* (Fig. 13.16).

Special types of coastal features develop in tropical and subtropical areas. These are **reefs**, made up of corals, algae and other marine invertebrates. The environment in which reefs can grow is very restricted in terms of salinity, temperature, sunlight, turbidity of water, etc. Common types of reef are shown in Fig. 13.17.

Glaciers are melting at the present time thus sea level is rising and submerging most shorelines—producing large **estuaries**. Some shores in tectonically-active areas, such as California, are emerging from the ocean, and one can see old wave-cut terraces in cliffs above the present beach.

Coasts can be grouped into three classes based on **tidal range:** less than two meters, 2-4 meters, and over 4 meters vertical range. Coasts with small tidal ranges typically are dominated by wave energy; those with large ranges are dominated by tidal current energy.

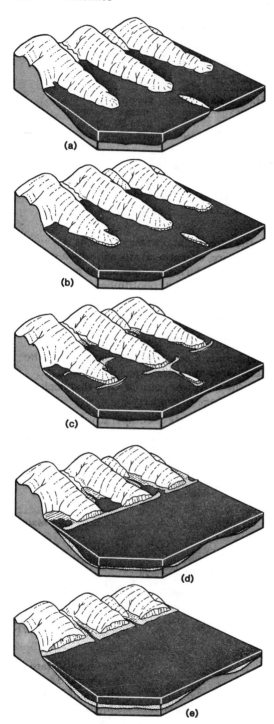

(a)

(b)

(c)

(d)

(e)

Figure 13.11. Cycle of coastal erosion: Sequence of sculptured forms developed through time. (A) Coast at start, with headlands, islands, and deep bays, (B) Headlands cliffed by surf, (C) Beaches, spits, and a tombolo are built as cliffing continues, (D) Cliffing reduces headlands to short stumps; spits join to form bay barriers, (E) Headlands and bays are eliminated; shoreline has become nearly straight. (From *Physical Geology*, 2nd Ed., by R. F. Flint and B. J. Skinner. John Wiley and Sons, Inc., New York, Copyright 1977.)

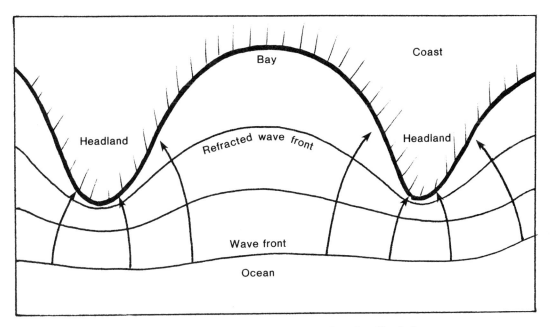

Figure 13.12. Diagram showing wave energy concentrated on headlands by refraction.

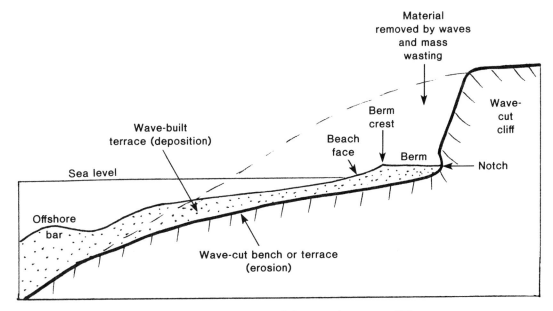

Figure 13.13. Shore profile consisting of erosional features (wave-cut cliff, wave-cut bench or terrace) and depositional features (wave-built terrace, bar, and berm).

Figure 13.14. Sea stacks, isolated from the headlands by wave erosion, western Ireland. (Photo courtesy Reinhard K. Frohlich.)

Figure 13.15. Sea arches produced by wave erosion of lava flows. Hawaii Volcanoes National Park. (Photo courtesy U. S. Department of the Interior, National Park Service.)

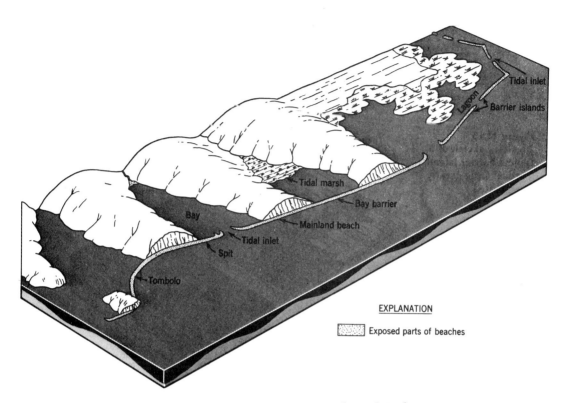

Figure 13.16. Types of depositional shore features. (From *Physical Geology*, 2nd Ed., by R. F. Flint and B. J. Skinner. John Wiley and Sons, New York, Copyright 1977.)

Coastlines can be classified with respect to plate boundaries. Inman and Nordstrom recognize leading edge or **collision coasts**, formed where two plates converge (continental or island-arc collision); **trailing edge coasts** where a coast faces a spreading ridge; **marginal sea coasts.** (Marginal seas, or basins, are tectonically related and typically are located between island arcs and continents, as along the western Pacific margin; other marginal seas occur between continents—the Mediterranean and Gulf of Mexico, for example.)

Coastlines also can be classified as **primary** if their shape is due to terrestrial (land) processes. Typically, the features are geologically young and the ocean as yet has not much affected them (fiords and deltas are two examples). **Secondary** coastlines show major effects of marine processes or organisms—barrier islands, coral-reefs, etc.

Another grouping would be into coasts of **construction** (deposition)—beaches, deltas, spits, reefs, dunes; or coasts of **destruction** (erosion)—cliffs, arches, caves, wave-cut benches. (Depositional features occur mostly on trailing edge and marginal sea coasts.)

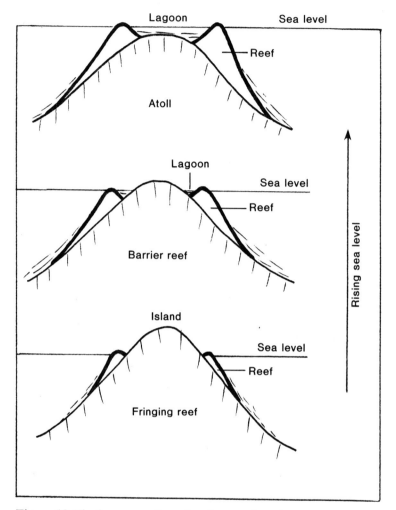

Figure 13.17. Sequence of coral reef types, from fringing to atoll, produced as sea level rises slowly.

14. Oceans

The oceans, together with the atmosphere, profoundly influence the earth's climate and constitute the major part of the hydrologic cycle. (The branch of the earth sciences that applies biology, chemistry, geology, and physics to the ocean environment is called *oceanography*.) Figures 1.2, 1.3, 1.4, and 1.11 show oceanographic features.

Oceans cover 60% of the northern hemisphere and 80% of the southern. The major topographic features of the oceans (introduced on p. 6) include 60,000 km-long *ridges*, *trenches* 40-120 km wide and 7-11 km deep, *abyssal hills* and *plains, seamounts* (probably 15-20,000 of them) and the *continental margins* (shelf, slope and rise). Some of these features are shown in Fig. 14.1. Fig. 14.2 is a topographic profile across the Atlantic Ocean basin. Ocean basins typically are 4-6 km below sea level. Figure 14.3 is a generalized cross-section of a passive (trailing) continental margin (e.g., the Atlantic, as contrasted with the active, plate collision-type margins of the Pacific). Note that continental crust extends seaward under the shelf and slope. Worldwide, the continental shelves average some 70 km in width and the break from shelf to slope occurs at an average depth of less than 150 meters. Some 20,000 years ago, sea level was at that break, reflecting glacial advance. Other causes of worldwide (eustatic) changes in sea level include sea-floor spreading, sedimentation rates, addition of volcanic water, etc.

Before 1960 it was generally accepted that ocean basins were permanent features and hence it should be possible to find in them a continuous record of sedimentation covering several billion years. Since then, through the new perspectives of plate tectonics, it has been recognized that the existing ocean basins are less than 200 million years old. Ocean crust is generally basaltic and constitutes part of a so-called *ophiolite suite* (which also includes some upper mantle material). A generalized ophiolite suite is shown in Fig. 14.4. The recognition of the phenomenon of sea-floor spreading (p. 209) resulted in the concept of plate tectonics and revolutionized the earth sciences (see Figs. 1.10, 3.19, 18.4).

Ocean Currents (Waves, including tides, are discussed in Section 13)

Ocean currents at the surface are largely wind-driven and move in large cells or *gyres*. In the northern hemisphere the motion is generally clockwise; movements are counterclockwise in the southern hemisphere. In a crude way these motions correspond to the circulation patterns of near-surface winds (Fig. 12.4) and illustrate the **Coriolis effect**: As the earth rotates, objects at the equator must have greater velocity than those at the poles (as a person skating in a circle at one end of a rope must travel faster than the person holding the other end). The earth rotates from west to east thus objects moving north from the equator will shift to the right (east) because of their higher velocity (and to the left, moving to the south).

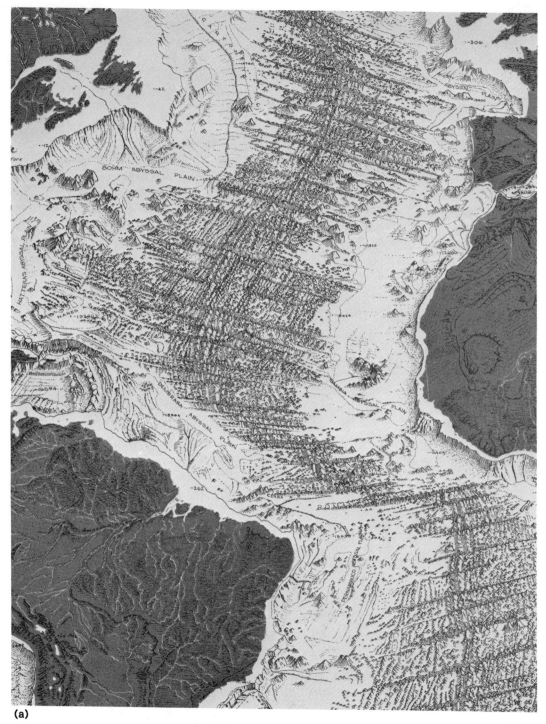

(a)

Figure 14.1. Features of the ocean basins. (A) Atlantic (cf., Fig. 1.4), (B) Western Pacific, (C) Indian. (Courtesy, Geological Society of America.)

Figure 14.1. continued

(b)

Figure 14.1. continued

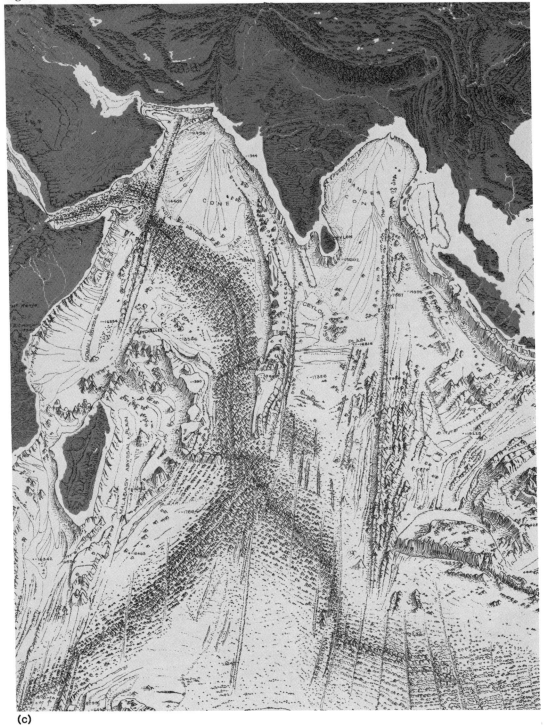

(c)

Figure 14.2. Generalized topographic profile across the Atlantic Ocean basin. (Courtesy, ALCOA.)

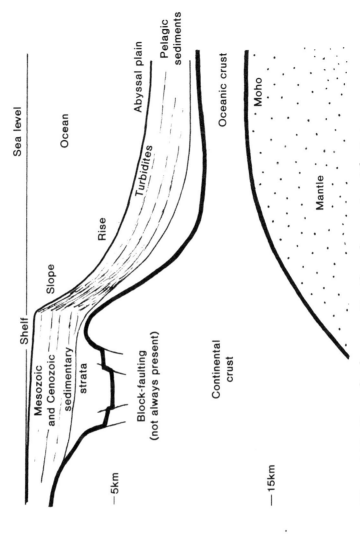

Figure 14.3. Generalized cross-section through a passive, Atlantic-type continental margin (contrasts with the active, Pacific-type in not having a trench).

167

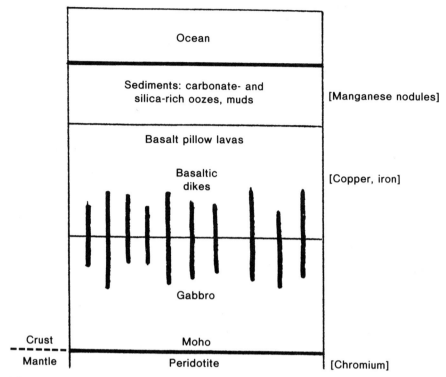

Figure 14.4. Generalized cross-section of an ophiolite suite (see also, Fig. 3.19).

Where cold subsurface water comes to the surface (*upwelling*) to take the place of water moved away by surface currents, nutrients are carried to biologically-active depths, giving increased productivity.

Deep-water movements are due to density differences caused by: increased *salinity* (total amount of dissolved solids in sea water) due to evaporation; *temperature variations* (which may be used to drive turbines); or *suspended sediment* (**turbidity currents**).

On relatively steep slopes and where there is a return flow (away from the coast) of water from major storm surges, turbidity currents may cut across the continental shelf and slope forming submarine canyons (some submarine canyons may have formed by river erosion when sea level was lower; others may have originated as tributary valleys flowing into the widening rift during early plate separation). Turbidity currents also deposit sediment (turbidites) as abyssal fans at the mouths of canyons, as well as in widespread graded beds on the continental rises (Fig. 14.1). Rises account for some 40% of the oceanic sediments.

Ocean Sediments

The composition of sea water remains fairly constant because the huge influx of material from rivers (p. 103) and volcanic processes is balanced by precipitation of sediment and secretion by organisms. Salinity averages 35 parts per thousand in the open ocean, primarily chloride, sodium, sulfate, magnesium, calcium, and potassium ions (see also, p. 90). Sediments in the oceans include

1. **Terrigenous material**—from weathering and decomposition of rocks, primarily on continents. Minerals are mainly silicates—quartz, clays, feldspars (carried by turbidity currents, ice-rafting, wind, etc.).
2. **Biogenic material**—derived from marine organisms: teeth, shells, bones, etc. Minerals include opal, calcite, apatite.
3. Material formed by **reactions between the sediments and sea water.** Manganese nodules are an example of precipitation around mineral fragments. Nodules have attracted much attention as a potential source of several scarce metals such as copper, nickel, cobalt, as well as manganese. Each nodule forms in roughly concentric layers at generally slow rates, 0.1 cm to over 100 cm per million years in some cases; they are extremely widespread on the ocean floor.

Characteristic deep-sea (**pelagic**) sediments are primarily red or brown **clays** and **calcium carbonate oozes** (shells of foraminifera, coccoliths) or **siliceous oozes** (shells of diatoms, radiolaria, silicoflagellates, sponge spicules). Carbonate oozes do not occur below the carbonate compensation depth (p. 128). Deep-sea sediment accumulates at rates of approximately 0.5 cm per thousand years whereas sediments on the shelves are deposited at rates of some 25 cm per thousand years.

Keeping in mind the hydrologic cycle, it will be clear that the oceans must be affected by materials transported, for example, by rivers and groundwater. Wastes and pollutants thus carried to the marine environment may become concentrated to toxic levels in particular types of sediment (e.g., clays) and/or organisms.

15. Man as a Geologic Agent

Man is a potent geologic agent. Society modifies the landscape, alters river patterns and shorelines, changes the atmosphere and groundwater compositions, influences climate and the biosphere, and in numerous other ways changes the rates at which natural processes operate.

Through farming, mining (Fig. 15.1), construction and related activities, society moves more materials than any other geologic agent. Removal of trees, urbanization, poor agricultural practices, etc., cause increased surface water runoff, with concomitant flooding, erosion, sedimentation in rivers, lowered water tables. Similarly, mass wasting can be exacerbated by man's activities which produce, for example, landslides, mudflows, and collapse of mining waste-piles.

Disposal of various wastes can cause groundwater pollution as well as the more obvious river contamination; these wastes also lead to pollution of the coastal, lake and ocean environments. Liquid wastes may contain potentially toxic metals, pesticides, radioactive elements, salt, etc. (In addition, deep injection of liquid wastes into disposal wells has caused earthquakes.) Solid wastes provide a longer-term source of pollution than do liquid wastes as they are leached slowly by circulating groundwater, ocean waters, etc.

Radioactive wastes pose a special threat because of the long half-lives of some of the decay products; some must be isolated from the environment for many thousands of years.

Atmospheric pollution includes heat, particulate (small solids) matter, and gases, such as oxides of sulfur and nitrogen, some of which are carcinogenic. Effects of atmospheric pollution also include climatic modification and more-acid rainfall, which in turn affect weathering rates, vegetation, etc.

Subsidence of the ground results from excessive withdrawal of groundwater, subsurface mining, petroleum drilling and similar activities, and causes, for example, collapse of structures and rupture of utility lines.

Other examples of how man is changing the natural environment will come immediately to mind. (A thorough treatment of man's interactions with the environment is given in courses such as Environmental Geology.) Each of us must heed the cliché that if we are not part of the solution to these environmental problems we are, in fact, part of the problem.

Figure 15.1. Butte Copper mine, Montana. (Photo courtesy Jon C. Boothroyd.)

INTERNAL PROCESSES

Earthquake of 1954, Fairview Park, Nevada; vertical fault movement. (Photo, National Oceanic and Atmospheric Administration, Environmental Data Service.)

Earthquake of 1964, Anchorage, Alaska; vertical ground displacement and building damage. (Photo, National Oceanic and Atmospheric Administration, Environmental Data Service.)

Earthquake of 1964, Niigata, Japan; aerial view of leaning apartment houses. (Photo, National Oceanic and Atmospheric Administration, Environmental Data Service.)

Internal Processes

Having examined the materials that make up the earth, and the external processes that shape its surface features, we turn now to internal processes. Obviously, in introducing the tectonic cycle (Fig. 1.10) and discussing igneous and metamorphic rocks, for example, we also were considering internal processes, thus some of the ideas presented here should be familiar. This section concludes with a restatement of the concept of plate tectonics and a brief look at the other planets to provide an introduction to, and framework within which to examine, the history of the earth (*Part II*).

16. Deformation

Deformation results from **stress** which, like pressure, is force per unit area. **Strain** is produced by stress and involves change in volume and/or shape. *Shear* produces change in shape without change in volume (adjacent parts of a solid slide past one another parallel to the plane of contact, like a deck of cards pushed sideways). *Compression* and *tension* produce a change in volume, size, and usually, shape. Whether a rock will *fold* or *fault* (**rupture**) depends on its strength, the rate at which it is deformed, the temperature, and the pressure. In general, folds occur at depth while faulting is more common near the surface (cf. plastic and brittle behavior in glaciers). Figure 16.1 shows relationships between stress and strain.

Plate boundaries are obvious environments in which rocks may be subjected to stresses. Plate collisions (subduction zones) which cause oceanic volcanic island arcs as well as folded mountain ranges on continents, produce compressional deformation features such as folds (Figs. 16.4 and 16.5) and reverse- and thrust-faults (Fig. 16.15). Faulting produces earthquakes when rupture is sudden. Orogeny, or mountain building, involves the compression of thick, geosynclinal accumulations of sediment (Fig. 16.23); accompanying the deformation is metamorphism, and intrusion of peridotites and granitic batholiths. (Later normal faulting and intrusion of dikes occur when there is a relaxation of compressional stress. The area is then subjected to extensive erosion, as it is topographically high and rivers actively downcut toward base level.) Plate separations (ocean-ridge spreading centers) show tensional rifting and normal faults. Where plates slide past each other, transform faults result (Fig. 16.18). Similar horizontal movements not part of moving plates are called transcurrent or strike-slip faults (Fig. 16.17). Figures 1.3 and 1.4 show tectonic features.

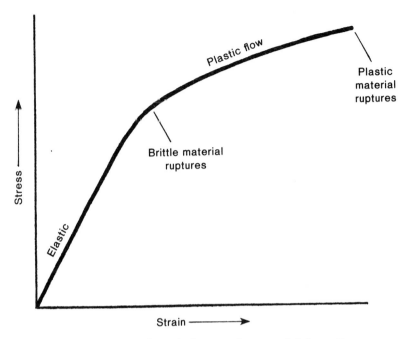

Figure 16.1. Generalized graph showing the type of deformation produced by increasing stress.

Folds

That layers of rock fold is shown in Fig. 16.2 (large scale) and Fig. 16.3 (smaller scale). Simple compression, dominantly in one direction, would give **symmetrical** folds (**anticlines** and **synclines**); continued compression would yield **isoclinal** folds, as shown in Fig. 16.4. If the compression had a shear component then **asymmetrical** anticlines and synclines would be produced (Fig. 16.5). As shown, continued compression would give folds with an **overturned** limb, and further deformation would produce a **recumbent fold**, part of which might become sheared off in a thrust fault, forming a **nappe structure.**

Note that in these illustrations, the dashed line, which represents the **axial plane** of the folds (dividing the fold in half), becomes more nearly horizontal as the shear component increases. Note also that different rock-types respond differently; consequently, an interbedded sandstone and shale sequence might show the shales thinned by squeezing in the limbs, and thickened in the crests of the folds; the more brittle (less ductile, more competent) sandstone would retain a more uniform thickness but would show cracks at the crests.

In the field, geologists measure the attitude of rock layers to determine the structure (i.e., the type of fold or fault). The basic measurements of attitude are called **strike** and **dip** (Fig. 16.6). On a map, the strike and dip of the bed in Fig. 16.6 would be shown ∠45 (striking N 30°E, dipping 45° to the S.E.). Vertical beds are symbolized by +, while horizontal units are symbolized by ⊕. Criteria for recognizing tops of beds are given on p. 67.

Figure 16.2. Folded sedimentary strata. Glenwood Canyon, Colorado. Rock layers in the left-hand part of the photo are close to vertical, those in the right-hand part are close to horizontal. (Photo courtesy Gerald J. Daub.)

Figure 16.3. Folded metamorphic rocks (light layers are marble, dark material is amphibolite). West central Vermont. (Photo courtesy Jon C. Boothroyd.)

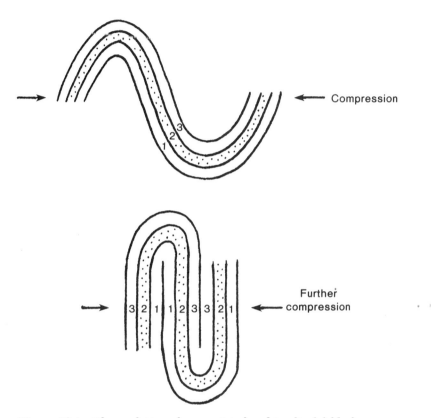

Figure 16.4. The evolution of symmetrical and isoclinal folds from originally horizontal beds.

Figure 16.7 is a map of a simple anticline and syncline where the ground surface has been planed by erosion. Figure 16.8 shows one side of an eroded fold with thin resistant layers separated by thicker, less-resistant units.

The oldest rock unit is in the center of an eroded anticline; the youngest unit is in the center of a syncline. In the former case, the beds dip out from the center, in the latter, they dip in toward the center. The distinction between anticline and syncline has some very important economic and environmental ramifications: because anticlines have provided traps for some 80% of the world's oil, they are much sought in petroleum exploration. Also, owing to the inward dip of a syncline's limbs, wastes put into the ground in one limb of a permeable

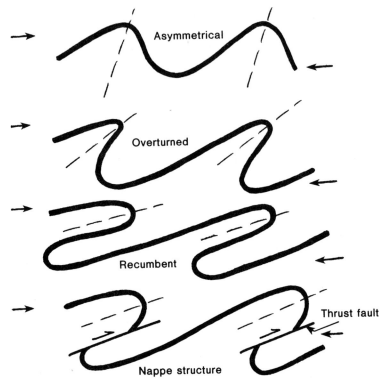

Figure 16.5. Sequence showing the evolution through asymmetrical folds, overturned folds, recumbent folds, and recumbent folds with thrust faults (nappe structure). Dashed line marks the axial plane.

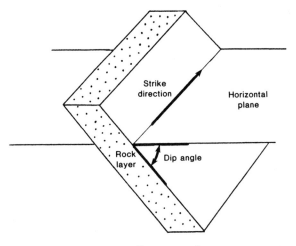

Figure 16.6. Diagram illustrating the measurement of strike and dip to establish the spatial attitude of a rock layer or other planar feature.

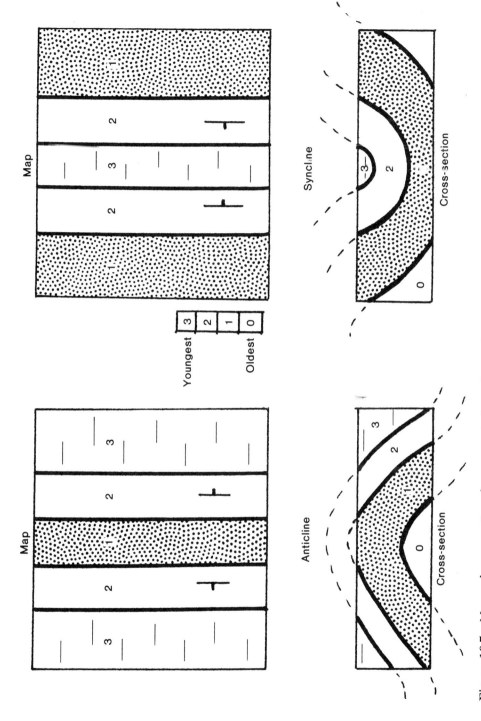

Figure 16.7. Map and cross-section of a symmetrical anticline and syncline eroded to a flat topographic surface. The youngest layer is in the center of a syncline, the oldest in the center of an anticline. (Note also that the dip symbols face in opposite directions.)

179

Figure 16.8. Part of an eroded fold (beds dip from lower right to upper left). The more easily-eroded layers are occupied by the reservoir; the resistant layers are vegetated. Front Range, Colorado. (Photo courtesy Gerald J. Daub.) See also, Fig. 16.24D.

unit, sandwiched between impermeable units, may be withdrawn from wells sunk into the other limb of the fold (Fig. 16.7). In an anticline this would not be so because the limbs dip away from each other (but contamination might occur off the map where the limb becomes part of an adjacent syncline).

Axes of folds usually are not horizontal over large distances but plunge into the ground giving characteristic V-shaped outcrop patterns on horizontal surfaces (Fig. 16.9).

Varieties of anticlines and synclines are **domes** and **basins**; they are shown in map view in Fig. 16.10. The state of Michigan is a large structural basin; the Black Hills are a large structural dome. A small dome, cut by erosion, is shown in Fig. 16.11. Figure 16.12 is of an eroded dome showing the concentric pattern of upturned beds. Domes can be produced by igneous intrusions, uprise of salt plugs or diapirs (Fig. 6.4), etc.

Monoclines are one-sided folds that often indicate a fault a depth (Figs. 16.13 and 16.14).

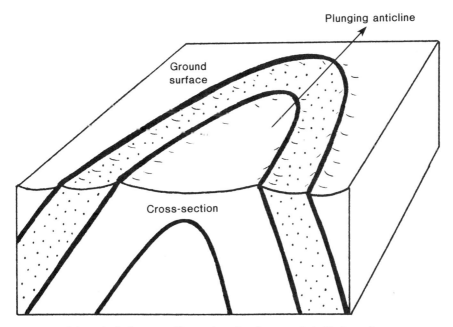

Figure 16.9. Block diagram illustrating the characteristic V-shaped outcrop patterns of an eroded plunging fold (anticline).

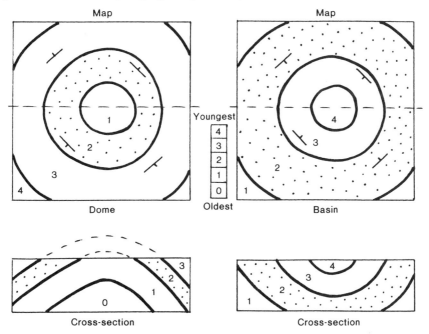

Figure 16.10. Map and cross-section of a dome and basin eroded to a flat topographic surface. The youngest rock unit is in the center of a basin, the oldest in the center of a dome.

Figure 16.11. Dome breached vertically, exposing the upwarped strata. Front Range, Colorado. (Photo courtesy Gerald J. Daub.)

Figure 16.12. Dome eroded to expose concentric layers of rock. Structure is 2 km across and 500 meters deep. Canyonlands National Park, Utah. (Photo courtesy U. S. Department of the Interior, National Park Service.)

183

Figure 16.13. Deeply-dissected monocline. Beds are horizontal in the left half of the photograph, and dip down in the right half (cf., Fig. 16.14). Capitol Reef National Monument. (Photo courtesy U. S. Department of the Interior, National Park Service.)

Faults

Faults are fractures in the earth's crust where one side has moved relative to the other (*joints* show no relative displacement). Faults produced under compression (crustal shortening) are called **reverse** and **thrust** faults; **normal** faults result from tension or crustal extension (Fig. 16.15). Where several normal faults are roughly parallel to each other, block faulting (including development of **horsts** and **grabens**) occurs, as shown in Fig. 16.16. Another type of fault, produced when motion is essentially horizontal (Fig. 16.17), is called **transcurrent, transverse, lateral,** or **strike-slip**. A special variety of lateral fault is the **transform** fault, recognized in 1965 by J. Tuzo Wilson; it involves horizontal slippage of lithospheric plates (Fig. 16.18). Note that normal and reverse faults involve mainly vertical movements; whereas in thrust, transcurrent and transform faults, movement of the rock units is primarily horizontal. Normal, reverse, and thrust faults are sometimes called **dip-slip** faults because, if we think of a fault plane instead of a bed in Fig. 16.6, it is clear that movement of the crustal blocks is up and down in the plane of the fault (in other words, in the direction in which the fault plane dips). Strike-slip fault movement is parallel to the strike of the fault plane; i.e., horizontal.

Movement along faults is by cumulative small increments over long periods of repeated reactivation. The Great Glen fault in Scotland shows strike-slip movement of approximately 100 km; evidence suggests the San Andreas fault has moved six times that much. Vertical fault movements are shown on page 173. Slow slip along faults (less than 2 cm/yr) commonly does not produce earthquakes.

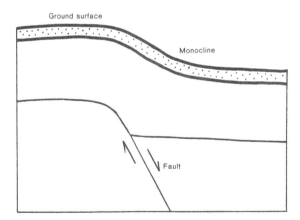

Figure 16.14. Diagrammatic cross-section of a monocline. Many, but not all, reflect a fault at depth.

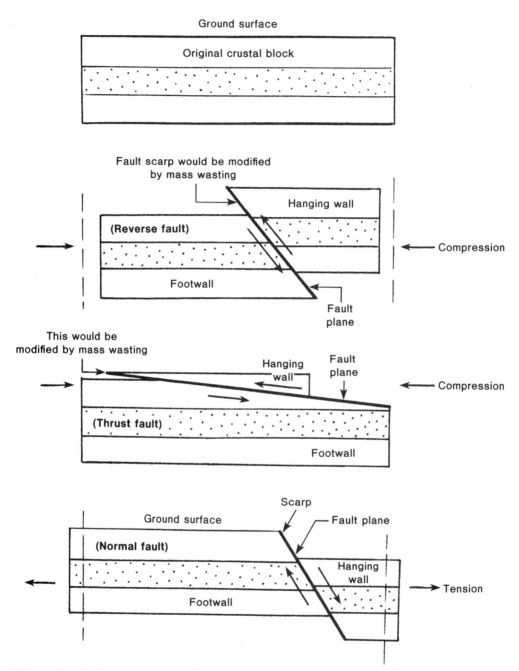

Figure 16.15. Common types of faults: compression produces crustal shortening and reverse faults, or low-angled varieties called thrust faults; tension produces crustal expansion and normal faults.

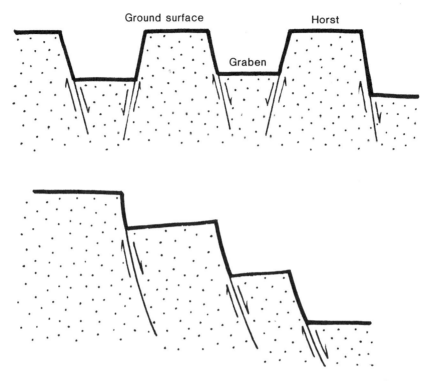

Figure 16.16. Step-faults, horsts and grabens produced by normal faulting.

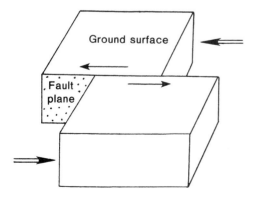

Figure 16.17. Diagram showing the type of movement in a strike-slip fault (also called lateral or transcurrent fault). Motion is essentially horizontal, in contrast to the largely vertical movements of reverse and normal faults.

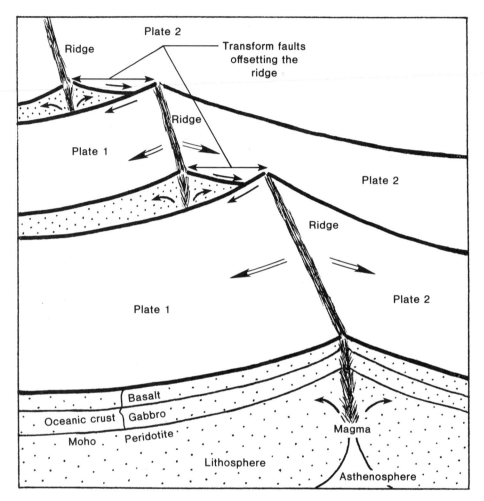

Figure 16.18. Diagrammatic view of transform faults offsetting an oceanic spreading ridge (divergent plate boundary). The transform fault is between the offset ridges; it differs from an ordinary strike-slip fault because the material being faulted is itself moving. Arrows on different sides of the fault show material moving in an opposite sense to the direction of ridge offset.

Unconformities

When erosion acts to plane off an uplifted or deformed area and sediment is then deposited over the erosional surface, an *unconformity* is produced (Figs. 16.19 and 16.20). In the geologic record, various types of unconformity show that rocks of certain ages are missing because of erosion and/or non-deposition; e.g., ***angular unconformity*** (Figs. 16.19 and 16.20), **disconformity** (horizontal beds above and below). Major unconformities can be traced as breaks in the geologic record across the North American craton. L. L. Sloss has used these regional unconformities to subdivide the sedimentary rock record of the past 600 million years into six "Sequences" of deposition, each separated from the next by a continent-wide erosional episode (the present situation). Many of the rock "Systems" of the relative geologic time scale (Cambrian, Ordovician, etc., shown in Table 4.4) were originally delimited by unconformities in their British and European type-sections (see *Part II*).

Figure 16.19. Angular unconformity. Rock layers at the top (with trees) are horizontal; more prominent layers underneath dip to the left (cf., Fig. 16.20). Dubois, Wind River Range, Wyoming. (Photo courtesy O. Don Hermes.)

Orogeny

As noted earlier, forces within the lithosphere and asthenosphere can deform large areas producing complex folding and faulting as part of plate collision (mountain-building or orogeny). (Also, simple vertical movements may occur without much deformation, a process called **epeirogeny**). Orogeny begins with the subsidence of linear troughs or **geosynclines** in which sediment accumulates. These troughs can develop along plate boundaries during the early stages of plate separation. Also during the early stages of plate separation, linear graben-like troughs, called **aulacogens**, form (present-day example is the East African Rift System). Aulacogens differ from geosynclines in being within a plate rather than along a plate margin. During plate collision, accumulated geosynclinal sediments at plate margins are compressed and deformed into mountain belts. Metamorphism and the intrusion of granitic batholiths accompany this compression and deformation. This is orogeny. The Sunda Islands (Sumatra, Java, Timor, etc.) are cited as an example of a modern geosynclinal and orogenic belt. The global extent of earlier geosynclines is shown in Fig. 16.21.

The evolution of the Appalachian foldbelt illustrates some of the interrelationships among various aspects of orogeny. The opening and closing of an ocean, an essential part of this evolution, is called a **Wilson cycle** and is illustrated in Fig. 16.22. Geosynclinal sediments on an Atlantic-type continental shelf and rise, collect in what Robert Dietz calls the **miogeocline** and **eugeocline,** respectively (Fig. 16.23). (They differ from Pacific-type geosynclines in being on passive [trailing] continental margins rather than on active [colliding] margins.) When the tectonic cycle changes the plate pattern, former passive margins can be brought together in collision, forming an orogenic folded mountain chain, as in Fig. 16.23. (Orogenic mountains owe much of their current rugged splendor to recent, essentially vertical, epeirogenic uplift or upwarping of the crust, followed by active erosion and dissection.)

Major types of mountains commonly not caused by orogeny include volcanic cones, often occurring in linear belts, fault-block mountains, erosional mountains, domal uplifts, etc. (Fig. 16.24). Elevated regions in the world are shown in Fig. 1.2.

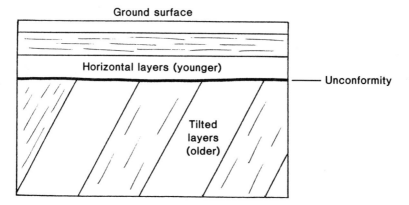

Figure 16.20. Diagram showing an angular unconformity (evidence of deposition, followed by deformation, followed by erosion, followed by deposition of horizontal beds).

Figure 16.21. Location of Paleozoic and Triassic geosynclines plotted on a reconstruction of the continents. The circled landmasses are five ancestral continents that existed during the Early Paleozoic before they formed Pangaea some 250 million years ago. (From *Earth History and Plate Tectonics*, 2nd Edition, by Carl K. Seyfert and Leslie A. Sirkin. Harper and Row, Publishers, Inc. Copyright 1979.)

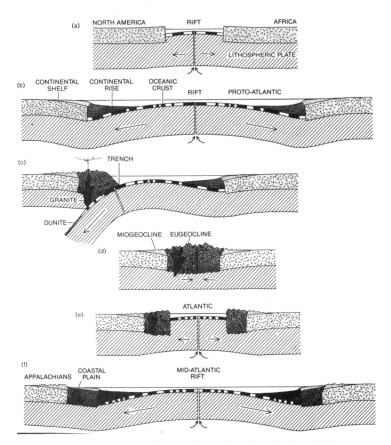

Figure 16.22. Mechanism of crumpling that produced the Appalachian foldbelt is depicted on the hypothesis that the Atlantic Ocean has opened, closed and reopened. In the late Precambrian (A), North America and Africa are split apart by a spreading rift, which inserts a new ocean basin. By the process of sea-floor spreading (B), the ancestral Atlantic Ocean opens. New oceanic crust is created as the plates on each side move apart. As the crust cools, its direction of magnetization takes the sign of the earth's magnetic field; the field periodically reverses, and the reversals are represented by the striped pattern. On the margin of each continent, sediments produce the geosynclinal couplet: miogeocline on the continental shelf, eugeocline on the ocean floor itself. The ancestral Atlantic now begins to close (C). The lithosphere breaks, forming a new plate boundary, and a trench is produced as lithosphere descends into the earth's mantle and is resorbed. The consequent underthrusting collapses the eugeocline, creating the ancient Appalachians. The eugeocline is intruded with ascending magmas that create plutons of granite and volcanic mountains of andesite. The proto-Atlantic is now fully closed (D). The opposing continental masses, each carrying a geosynclinal couplet, are sutured together, leaving only a transform fault (vertical black line). The shear contains squeezed-up pods of ultramafic mantle-rock. Sediments eroded from the mountain foldbelt create deltas and fluvial deposits collectively called molasse. North America and Africa were apparently joined in this way between 350 million and 225 million years ago. About 180 million years ago (E) the present Atlantic reopened near the old suture line. Today (F) the central North Atlantic is opening at the rate of three centimeters per year, creating new geosynclines. (From *"Geosynclines, Mountains and Continent-building"* by Robert S. Dietz, Copyright 1972 by Scientific American, Inc. All rights reserved.)

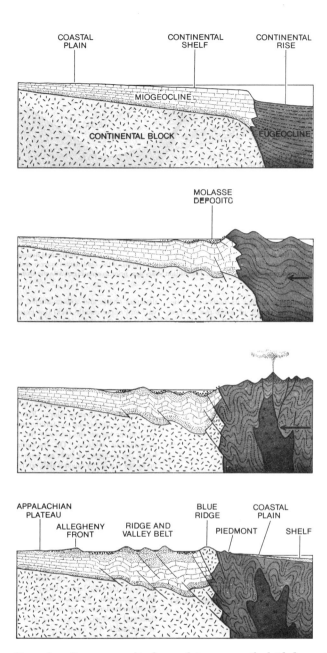

Figure 16.23. Crumpling of earlier geosynclinal couplet, apparently laid down in late Precambrian and early Paleozoic time more than 450 million years ago, produced the Appalachian foldbelt. The four-part sequence shows how the miogeocline, or western part of the geosynclinal couplet, was folded into the series of ridges between the Blue Ridge line and the Allegheny front. The eugeocline, altered by heat, pressure and volcanism, formed a lofty range of mountains, now almost completely eroded, east of the Blue Ridge line. Recent evidence suggests that thrust faulting has moved the Blue Ridge more than shown here; perhaps over 200 km. (From *"Geosynclines, Mountains and Continent-building"* by Robert S. Dietz, Copyright 1972 by Scientific American, Inc. All rights reserved.)

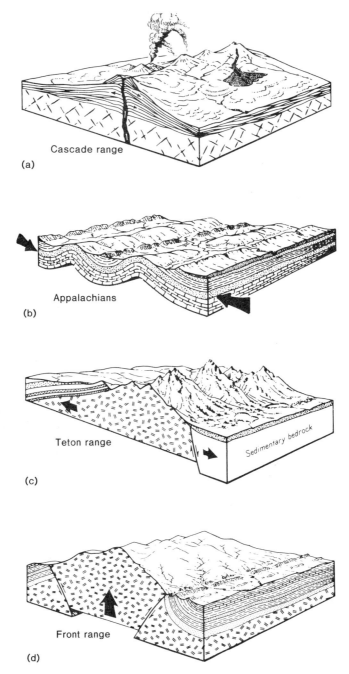

Figure 16.24. Types of mountains (A) mountains formed by volcanic action, (B) mountains resulting from folding layers of rock, (C) mountains formed from fault blocks, (D) mountains formed as a result of vertical uplift. (From U. S. Geological Survey.)

Isostasy

Isostasy is the concept of flotational balance, whereby high mountain ranges extend a root of crustal material (Fig. 16.25) into the denser mantle. In this way, greater thicknesses of less-dense rocks are in equilibrium with thinner units of more dense rock. The higher the mountain, the deeper the root. As erosion lowers the mountain, the root rises, much like an unloaded ship. Continental glaciation provides an example of isostasy: loading of the continent during glaciation depresses the crust farther into the mantle; the continent rebounds when the glaciers melt and retreat.

The idea that mountains had roots of material less dense than the surrounding mantle was proposed to explain an observation made around 1850 by surveyors in the Himalayas. Knowing that their plumb-line would be attracted by the mass of the nearby mountains (attraction is proportional to $\dfrac{\text{mass A} \times \text{mass B}}{(\text{distance between them})^2}$ surveyors made corrections in their calculations based on their estimate of the volume of the mountains and the density of the rocks in them. The observed attraction, however, was less than 1/5 the predicted amount. The explanation proposed by G. B. Airy, since verified by geophysical evidence, is that less-dense (crustal) material underlies each mountain (i.e., its root) thus the effective attraction is less than predicted because the total mass is less than if denser mantle material were near the surface. (J. H. Pratt proposed, in an alternative explanation, that mountains did not have roots but, rather, stood high because they were composed of rocks less dense than adjacent lowland areas. In the Airy model the continental crust is reasonably homogeneous; in the Pratt model the crust is not homogeneous and its base is horizontal.)

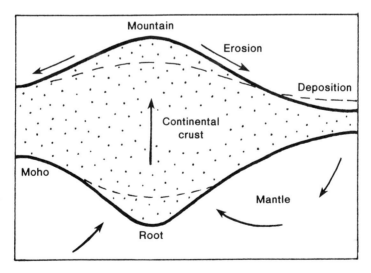

Figure 16.25. Isostasy. Mountains stand high because they are underlain by large roots of low-density crustal material (somewhat like an iceberg). As the mountain is eroded, the root rises (positions shown by the dashed lines).

17. Earthquakes

Some earthquakes are caused by the movement of magma underneath volcanoes (and can be used to predict eruptions). Most earthquakes, however, are tectonically related and occur along plate boundaries (especially during plate collision) where an obvious mechanism for deforming rocks exists. Tectonic earthquakes result from the sudden release of **elastic strain energy** that has been stored for tens or even hundreds of years in rocks undergoing deformation. When rock is deformed beyond its strength, an earthquake occurs due to rupturing or faulting of the rock, and **seismic** (earthquake) **waves** are produced. (The distribution of recent major earthquakes is shown in Fig. 17.1.) According to the **elastic rebound theory**, plates do not slide past each other smoothly but in a series of jerky movements; each jerk produces an earthquake. Often an earthquake is preceded by smaller shocks (**foreshocks**) over a period of perhaps weeks, and is followed by **aftershocks**, sometimes several hundred in number.

Figure 17.1. Major earthquakes (Magnitude 7.0 or greater) 1968-1974 (cf., Fig. 3.9). (Reported by the Center for Short-Lived Phenomena.)

Some earthquakes, however, do not have an obvious cause as they are not related to plate boundaries or volcanic eruption—the large Charleston, S. C. earthquake of 1886 and the even larger Missouri 'quakes of 1811 and 1812, for example. Eastern U. S. earthquake areas are shown in Fig. 17.2.

Most earthquakes originate (have their **focus**) at depths less than 30 km (in central California most are less than 5 km deep). Shallow-focus earthquakes are those originating within 70 km of the surface, deep-focus earthquakes have their origin deeper than 300 km (intermediate-depth earthquakes are in-between). Deep-focus (and most of the very large) earthquakes typically occur on the continental side of oceanic trenches (along the *Benioff zone*, which coincides with the subduction zone of plate tectonics, Figs. 1.11 and 5.1). The deepest earthquake focus recorded is 700 km (presumably elastic strain is not possible at depths greater than 700 km); possibly this represents the greatest depth at which earthquakes can occur as below this depth all subducted lithosphere has been absorbed into the mantle.

Earthquake Waves

Earthquake energy is transmitted as **body** (P and S) **waves** through the earth, and as slower **surface** (L) **waves** along the earth's surface; the latter include Love waves (shaking motion like horizontal S waves) and Rayleigh waves (like deep-water waves).

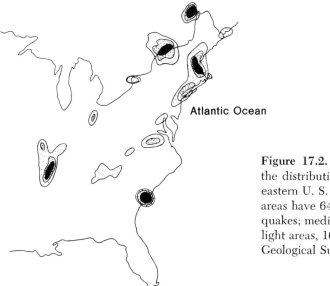

Atlantic Ocean

Figure 17.2. Simplified map showing the distribution of earthquakes in the eastern U. S. from 1800-1972. Darkest areas have 64 or more observed earthquakes; medium shading, 32 or more; light areas, 16 or more. (From U. S. Geological Survey.)

There are two types of body waves—compressional and transverse. **Compressional** (or push-pull, primary, or P) waves travel at velocities ranging from 5 to almost 14 km per sec at different depths in the earth; changes in the velocity are caused by materials having different elastic constants and densities. P waves travel through any type of material.

Transverse (or shear, shake, secondary, or S) waves travel only through materials that resist change in shape; thus they do not travel through liquids. This fact is of extreme importance as the loss of S waves in the earth's interior is evidence for a liquid outer core. P waves travel at about 1.75 the velocity of S waves. This means that the farther the waves travel from their origin, the greater the difference there will be between the time of arrival of P and S waves (Fig. 17.3). This time-difference is used to locate the earthquake **epicenter** (on the surface, directly above the focus) as shown in Fig. 17.4.

All three types of waves are recorded on a seismogram (Fig. 17.5) by a seismograph. The amplitude of the waves is a function of the amount of ground shaking, which in turn is a measure of the earthquake energy released. The **Richter scale** measures the **magnitude** of an earthquake using these instrumental records and is a quantitative scale based on the amplitude of the largest shock waves recorded on the seismogram. (Current practice expands the original Richter magnitude concept to allow for the fact that deep-focus earthquakes do not produce the large surface waves typical of shallow-focus earthquakes: magnitude is calculated from the amplitude of P waves [which are not affected by depth of focus] as well

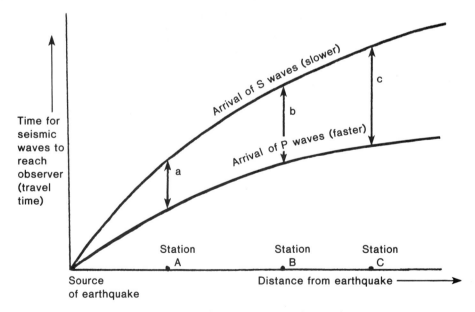

Figure 17.3. Generalized relationship between arrival times for seismic waves and distance from the earthquake epicenter. P waves travel faster than S waves thus the farther they travel the greater difference there will be in their times of arrival.

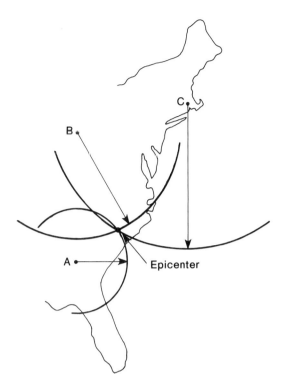

Figure 17.4. Locating an earthquake epicenter from the records at three seismic stations. Radius of each circle is the station's distance from the earthquake as calculated in Fig. 17.3. The epicenter is where the three circles intersect.

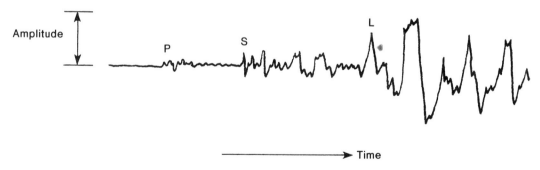

Figure 17.5. Simplified seismogram showing the general appearance of P, S and L waves.

as from the amplitude of surface waves.) Correction factors have been established through numerous observations so that all seismograms recording a given earthquake show the same magnitude regardless of distance from the focus (unlike intensity which is a qualitative measurement, expressed in Roman numerals, of damage and human responses, and which decreases away from the epicenter). The amplitudes of seismic waves are recorded on a logarithmic scale thus each increase in Richter magnitude number corresponds to a ten-fold increase in amplitude. This increase represents a 30-fold increase in the energy released by the earthquake. (A Richter magnitude 7 has 27,000 times the energy of a magnitude 4 [$30 \times 30 \times 30$]; one therefore can see the importance of the decimals used in the Richter scale). Some 20 earthquakes per year have magnitude 7 or greater, but perhaps 20,000 per year range between 4 and 7. An earthquake of magnitude 5 (intensity VI) releases energy roughly equivalent to that of a small atom bomb; one of magnitude 6 (intensity VII) is equivalent to a one megaton hydrogen bomb. Magnitudes greater than 8 are equivalent to tens of thousands of hydrogen bombs. The greatest magnitude recorded is 8.9 (1906, Colombia-Ecuador; 1933, Japan).

Damage

A non-instrumental way to describe the severity of an earthquake is to express its effects qualitatively in terms of damage and the responses of people affected. These effects indicate the **intensity** of an earthquake; criteria are summarized in Table 17.1. Intensity (unlike magnitude) will decrease as one moves away from the epicenter (although local variability is produced by the responses of different materials). Damage caused by earthquakes includes landslides, tsunami (seismic sea-waves that can travel over 800 km per hour), building-collapse with resultant fires, etc. Based on historical earthquake data it is possible to anticipate the shaking to be expected during an earthquake (Fig. 17.6), and hence the damage that might be caused. The values given are for shaking in solid bedrock; unconsolidated or poorly-consolidated material responds much more disastrously. Also, the types of man-made structures will obviously affect the extent of the damage; flexible wooden structures survive better than rigid brick ones, for example. Despite the huge loss of life and extreme damage caused by earthquakes and their ramifications, society continues to build communities on known earthquake-prone faults (the San Andreas Fault of California is the most publicized example). Figure 17.7 shows how deceptively innocuous the San Andreas Fault may appear. Recall, however, that earthquakes killed some 700,000 people in 1976 alone.

Prediction of Earthquakes

It is possible, to some extent, to predict earthquakes, using various criteria such as **seismic gaps.** Seismic gaps are those parts of active faults along which earthquakes have not occurred recently and where strain energy thus has not been released. These areas are potential earthquake locations.

Some precursors of earthquakes are outlined in Fig. 17.8. The warnings commonly used can be explained by the **dilatancy model**: As stress builds up over perhaps several years, min-

TABLE 17.1   Earthquake intensities (based on Modified Mercalli Scale)

Intensity	Description	Equivalent Magnitude on Richter Scale
I	Not felt; recorded by instruments	
II	Felt by some, especially on upper floors. Some suspended objects swing.	—3—
III	Noticeable indoors but not recognized by many. Vibration like a passing truck.	
IV	Some awakened. Like truck striking building. Dishes rattle.	—4—
V	Felt by most. Some windows broken.	
VI	Felt by all. Many run outdoors. Some heavy objects moved; minor fallen plaster and cracked chimneys. Slight damage.	—5—
VII	Most run outdoors. Noticeable in moving vehicles. Considerable damage to poorly-designed structures; negligible damage to well-designed and constructed buildings.	—6—
VIII	Slight damage to specially designed structures; severe damage to weak structures. Chimneys, walls, etc. fall. Heavy furniture overturned.	
IX	Considerable damage even to specially designed structures. Buildings moved off foundations. Ground cracks. Underground pipes broken.	—7—
X	Most masonry and frame structures and foundations destroyed. Ground severely cracked. Landslides. Rails are bent.	
XI	Bridges destroyed. Fissures open in the ground. Few masonry structures survive.	—8—
XII	Total damage. Objects thrown into the air. Wave motions are seen on the ground surface.	

ute cracks open in the rocks and they dilate. Water in the existing rock-openings (pores) thus has additional cracks into which it can move and the fluid pore pressure drops. This strengthens the rock. Measurements of electrical conductivity and the velocity of P waves show an initial decrease (because of the lowered fluid pressure) followed by an increase as the openings fill with water moving in from nearby areas. As pore pressures rises again, conductivity and P wave velocity increase. The increase in pore pressure weakens the rock, leading to rupture and an earthquake. The longer it takes for the pore pressure to return to normal (because of a greater number of cracks), the larger the earthquake will be. The detection and measurement of these phenomena mean it may be possible to predict not only when and where, but also how big an earthquake will be.

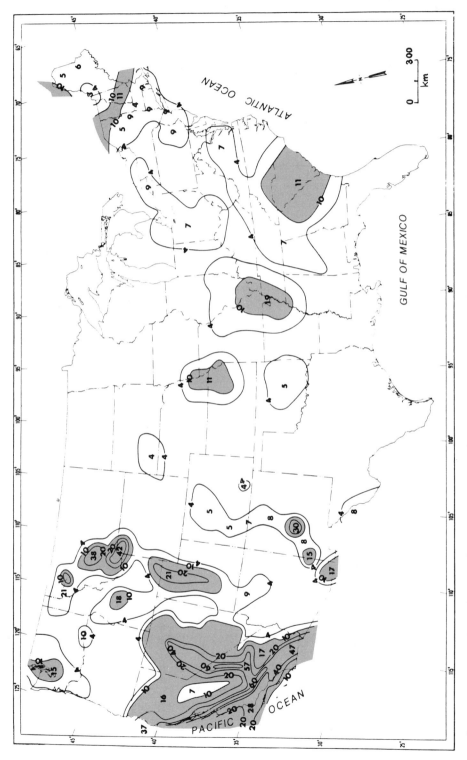

Figure 17.6. Map shows expectable levels of earthquake shaking hazards. Levels of ground shaking for different regions are shown by contour lines which express in percentages of the force of gravity the maximum amount of shaking likely to occur at least once in a 50-year period. Damage would occur where values are 10% or more. (From U. S. Geological Survey.)

Figure 17.7. Offset in a drainage ditch produced by movement along the San Andreas Fault, Hollister, California. Movement is 1.5 cm/yr. (Photo courtesy O. Don Hermes.)

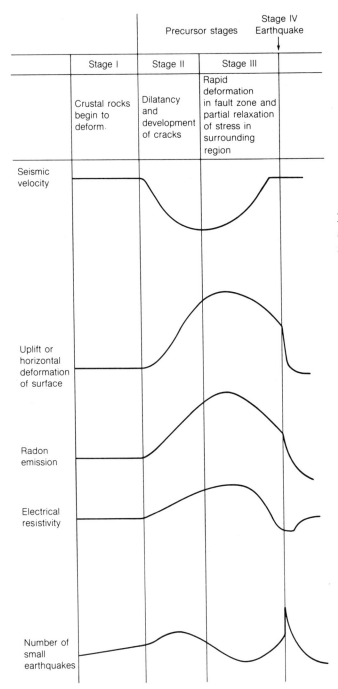

Figure 17.8. Earthquake precursors and their relation to dilatancy. When rocks are stressed by crustal forces, they are deformed elastically (stage I). Before faulting occurs, cracks open up, the rocks become dilatant and expand (stage II). This expansion is manifested as uplifting or tilting of the surface. The opening of cracks also results in a reduction in seismic velocity and an increase in the transfer of radioactive radon gas to water percolating through the rock. Small earthquakes increase in number. Stage II can serve for long-term predictions, years in advance of a great earthquake. In the third stage, rapid deformation occurs along portions of the fault that will break, relaxing the stress in the surrounding rocks, so that cracks close. Uplift or tilting would decrease or reverse in direction, seismic velocity would increase and small earthquakes would decrease in number. Days or hours before the fault breaks, the number of small earthquakes increases rapidly, signalling the imminence of the earthquake. (From *Earth*, Second Edition, by Frank Press and Raymond Siever. W. H. Freeman and Company, Copyright 1978.)

The discovery that fluids pumped to considerable depths can trigger small earthquakes (fluids pumped into an Army waste-disposal well near Denver led to this discovery) has held out some possibility that the accumulating strain energy in earthquake-prone faults could be released at safe levels by periodically initiating small earthquakes. The risk of causing unwanted larger earthquakes is certainly great but this method seems a promising research direction.

Earth's Interior

Earthquakes give us information about the earth's interior. Because direct observation of the interior is impossible, seismic interpretations are indispensible to the earth scientist; much of the information on the earth's interior and the tectonic cycle presented earlier in this book (e.g., Fig. 1.5) is based on seismic evidence. The following figures show some examples of seismic information. Figure 17.9 illustrates the change in velocity of seismic waves in the lithosphere and asthenosphere. Figures 17.10 and 17.11 show paths of P and S waves, respectively, through the earth; Fig. 17.12 shows a few of the additional paths that are possible.

Man-made seismic waves (explosions) are used in geophysical exploration for natural resources and to determine the structure of underlying rocks for major construction projects. Seismic waves which are reflected from various rock layers at depth are recorded at the surface. These seismic reflection records can show the depth to particular reflectors such as ore bodies, as well as delineating folds, faults and other potential oil traps.

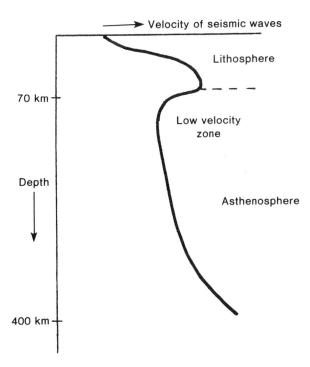

Figure 17.9. General changes in seismic (earthquake) wave velocity with depth. Note the abrupt decrease at the base of the lithosphere, marking the top of the low velocity layer in the asthenosphere.

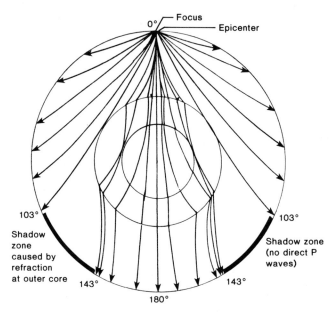

Figure 17.10. Simplified diagram showing paths of P waves through the earth.

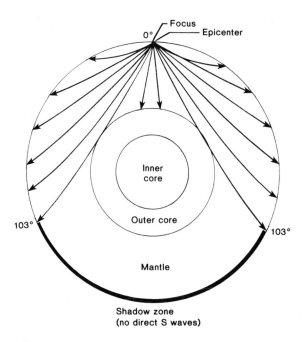

Figure 17.11. Simplified diagram showing paths of S waves through the earth.

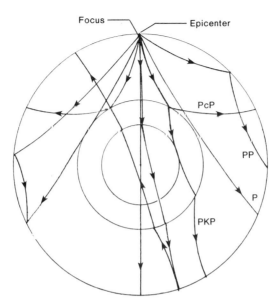

Figure 17.12. A few of the many reflected and refracted paths of P and S waves.

18. Plate Tectonics

There have been many attempts to explain the distribution of mountain ranges and other features through global hypotheses. One of the earliest was that the earth was contracting as it cooled (much like a dried apple). This contraction explained the prevalent compressional features seen on continents but did not explain the linearity of folded mountains (Fig. 1.3) or tensional features later found to be dominant in oceans. Also, it was proposed before the discovery of radioactivity in 1896 indicated a source of internal heat for the earth. An alternative idea was the expanding-earth hypothesis which explained the tensional features but was less successful with the compressional ones (folds, reverse- and thrust-faults).

A model was needed that would explain the distribution of all types of geologic features. That model is **plate tectonics** (proposed in 1967 by McKenzie and Parker and in 1968 by W. Jason Morgan) which incorporates earlier models of sea-floor spreading, continental drift, transform faults, etc. The essentials of plate tectonics have been given in the section on the tectonic cycle (e.g., Figs. 1.10 and 1.11) and referred to in discussions of igneous and metamorphic rocks, earthquakes, and elsewhere. Before summarizing the evidence for plate tectonics and the characteristics of plate boundaries, it is necessary to review some basic information about the **earth's magnetic field**, as the interpretation of magnetic properties in the rock record is an essential part of plate tectonics.

Magnetically, the earth behaves as if it contained a large bar (dipole) magnet lying almost parallel to the axis of the earth's rotation (the geographic poles). The angle between true (geographic) north and magnetic north is called the **declination**; this angle (east or west of true north) varies depending on the observer's location. The magnetic pole moves slowly around the rotational pole but remains close to it.

A magnetized needle not only points to magnetic north, its north-seeking end also is pulled toward the surface of the earth by the magnetic field (Fig. 18.1). The extent to which the needle is pulled down depends on the latitude at which the observation is made: at the magnetic poles the needle would be vertical, whereas at the equator the needle would be horizontal. The angle between the magnetic field and the earth's surface is called the **inclination**.

Throughout the geologic record the **polarity** of the earth's magnetic field has alternated between *normal* (present polarity) and *reversed* (as if the north magnetic pole were at the south pole). Extinction of some groups of organisms seems to correlate with polarity reversals but the cause is not certain (unshielded cosmic rays?).

These magnetic properties are preserved by the magnetization of minerals in rocks. The three major types of remanent magnetism are: (a) **thermoremanent**—where igneous rocks are

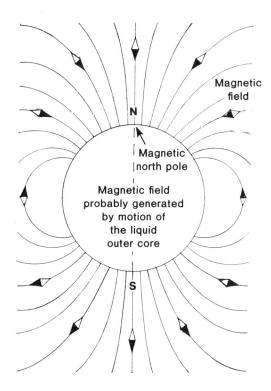

Figure 18.1. Simplified diagram of the earth's magnetic field. (The magnetic pole is about 11° from the rotational pole.)

magnetized in the earth's magnetic field as they crystallize from magma and cool below the *Curie point* (temperature, around 550°C, above which magnetization of a particular mineral type is not possible); (b) **depositional**—magnetic orientation of iron oxide sedimentary particles as they are deposited in water; (c) **chemical**—magnetization of diagenetic iron cement (e.g., hematite in sandstones). The study of ancient pole positions and polarities (**paleomagnetism**) is a key aspect of plate tectonics, as outlined in the following pages.

The Basic Assumptions of Plate Tectonics Are:

1. The outer part of the earth (lithosphere) behaves as a series of rigid slabs or plates (Figs. 18.2 and 1.11). There may be vertical movements within a plate, but major deformation generally takes place at plate boundaries.

2. Each plate is moving relative to the other plates (7 major and several smaller ones). The three types of plate boundary are *divergent* (ridge), *convergent* (trench, subduction), and *transform* (Table 1.1).

   Divergent boundaries occur where sub-lithospheric convection currents produce seafloor spreading with the extrusion of basaltic magma at the oceanic ridges. The basalt is derived from partial melting of peridotite in the upper mantle. In addition to adding to the earth's lithosphere, the basalt records the position of the earth's magnetic field at the time of

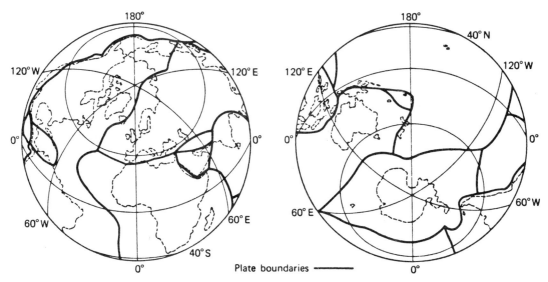

Figure 18.2. Plate boundaries (see also, Fig. 1.11). (From *The Way the Earth Works*, by Peter J. Wyllie. John Wiley and Sons, Inc., New York, Copyright 1976.)

extrusion as the lava becomes magnetized on cooling below the Curie temperature. Linear magnetic anomalies form where strips of basalt magnetized in the direction of the earth's present magnetic field (normal polarity) alternate with strips having reversed polarity. As recognized by Vine and Mathews in 1963, these strips preserve the history of spreading in a pattern symmetrical about the ridges, shown in Fig. 18.3. Numerous small earthquakes are produced during divergence and many metals are brought into the marine environment (as part of the new lithosphere) where they are concentrated in muds. Heat flow is high.

Convergent boundaries involve subduction as one plate descends under the other, presumably at the location of downward-moving, sub-lithospheric convection currents. Typical features produced include trenches, volcanic island arcs, and young, linear folded mountain belts (orogenic zones). The rock record is much more diverse and complex than at divergent boundaries. Characteristic boundary features depend on whether convergence involves two lithosphere plates carrying mainly oceanic crust (e.g., Japan—volcanic island arc, Fig. 3.18); one carrying oceanic crust and one carrying continental crust (e.g., the Andes—volcanic linear mountain belt, Fig. 3.17), or both plates carrying continental crust (e.g., Alps and Himalayas, linear non-volcanic mountain belts, Fig. 3.21). Continental igneous rocks tend to be andesites, rhyolites, and granites derived from partial melting of continental crustal material. Continental basalts also occur as a result of partial melting of deeper, subcrustal peridotites and subducted oceanic basalt. Regional metamorphism (with some local contact metamorphism), large, deep-focus earthquakes, and metal deposits of Andean- and island-arc types (Fig. 6.2) also are typical of convergence.

Transform faults, the third type of boundary (Fig. 16.18), permit plates to slide past each other along strike-slip faults. These faults can connect divergent and/or convergent

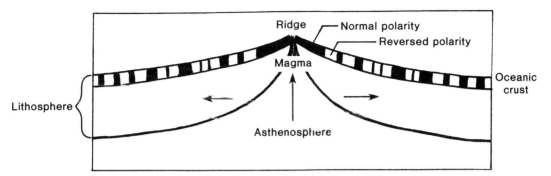

Figure 18.3. Cross-section showing the distribution of normal- and reversed-polarity lavas generated at a spreading ridge over approximately 5 million years. Note the patterns are symmetrical about the ridge.

boundaries and may involve major lateral movement. The San Andreas Fault of California is a rare example that occurs on land; typically the faults offset oceanic ridges (Fig. 1.4). Igneous activity is not associated with transform faults but severe, shallow-focus earthquakes are common between the offset ridges where the two plates move against each other. Because transform faults involve sideways-displacement of the elevated ridge, these faults may expose deeper crustal material (the ophiolite suite).

Rates of plate movement (measured by dating magnetic anomalies, for example) vary from 1 to 20 cm/yr (however, the East African Rift is separating at less than 0.1 cm/yr). These rates have been remarkably constant for each locality over the past tens of millions of years.

The plate tectonic model involves three interrelated types of movement: continental drift, sea-floor spreading, and polar wandering. **Continental drift** was first suggested when people looked at maps of Africa and South America and saw they could fit the coastlines together reasonably well (for example, Von Humboldt in 1801 and Snider-Pellegrini in 1858). The early concept of continental drift, expounded by Alfred Wegener in 1912, had the continents separating from a single landmass, called **Pangaea**, by moving across the ocean basins, much as ships sailing through the sea. In the current concept, continental drift results from the separation of lithospheric plates on which the continents are passive passengers. Fig. 18.4 shows the break-up of Pangaea. The continents do not move across the ocean basins; the ocean basins grow larger by addition of new lithosphere at divergent ridges. This is the concept of **sea-floor spreading** (proposed by Harry Hess in 1960) which obviously explains continental drift as well as characteristics of the ocean basins (symmetrical patterns of ages of sediments and lavas, magnetic polarities, etc.). **Polar wandering**, as discussed by Wegener, involved the hypothesis that either the earth's axis of rotation moves relative to the whole earth, or the outer part of the earth slides over the inner mantle and, hence, moves relative to the earth's rotational poles. Evidence that landmasses have not always occupied their present position with respect to the geographic poles involves interpretation of climates as recorded in rocks; i.e., ancient evaporites, coral reefs and other warm-water assemblages near the present poles, and old glacial deposits near the equator. A different line of evidence, not avail-

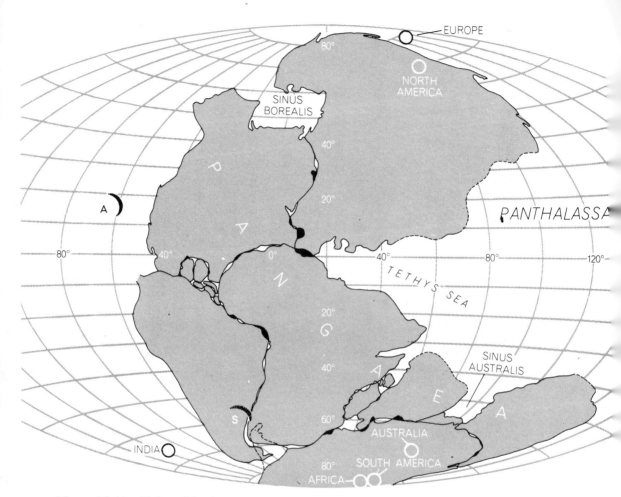

Figure 18.4A.   Universal land mass Pangaea may have looked like this 200
million years ago. Panthalassa was the ancestral Pacific Ocean. The Tethys Sea
(the ancestral Mediterranean) formed a large bay separating Africa and Eurasia.
The relative positions of the continents, except for India, are based on best fits
made by computer, using the 1,000-fathom isobath to define continental boundaries. When the continents are arranged as shown the relative locations of the
magnetic poles in Permian times are displaced to the positions marked by circles.
Ideally these positions should cluster near the geographic poles. The hatched
crescents (A and S) serve as modern geographic reference points; they represent
the Antilles arc in the West Indies and Scotia arc in the extreme South Atlantic.

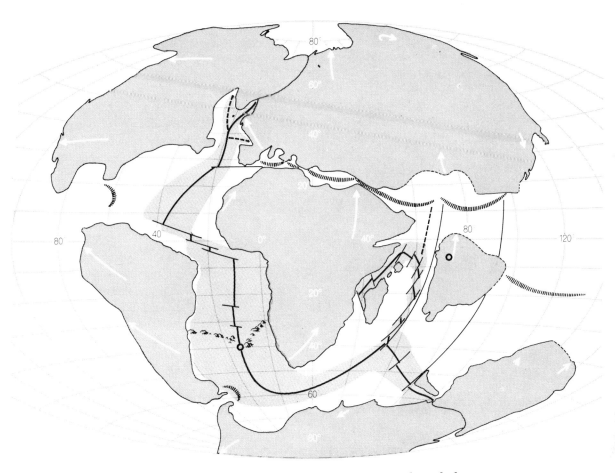

Figure 18.4B. After 135 million years of drift, 65 million years ago at the end of the Cretaceous period, the South Atlantic has widened into a major ocean. A new rift has carved Madagascar away from Africa. The rift in the North Atlantic has switched from the west side to the east side of Greenland. The Mediterranean Sea is clearly recognizable. Australia still remains attached to Antarctica. An extensive north-south trench (not shown) must also have existed in the Pacific to absorb the westward drift of the North American and South American plates. Note that the central meridian in this reconstruction is 20 degrees east of the Greenwich meridian. (From *"The Breakup of Pangaea,"* by Robert S. Dietz and John C. Holden, Copyright 1970 by Scientific American Inc. All rights reserved.)

able to Wegener, comes from the record of old *magnetic* pole positions preserved in rocks. These paleomagnetic data show that the magnetic poles for old rocks are in quite different positions from the present poles (magnetic pole and rotational or geographic pole are assumed to have remained close together as the earth's rotation is believed to generate the earth's di-pole magnetic field). All these observations can be explained by plate movements causing continents to rotate and/or move into different latitudes. The current view is that movement is primarily by the continental rocks (as part of lithospheric plates) and not by the poles, thus the term "polar wandering" is somewhat misleading. The term "apparent polar wandering" is used in paleomagnetism to indicate that the magnetized rocks themselves have moved.

Evidence for Plate Tectonics Includes:

1. Geographic fit of landmasses (e.g., Fig. 18.4).
2. Paleontological record of isolated occurrences of identical fossils on now-separated continents.
3. Glacial striations and other evidence of the direction of ice-movement some 275 million years ago show that for the present configuration of the southern continents, the ice (most improbably) advanced from the sea. More likely, the landmasses at that time formed a single continent, *Gondwanaland,* and the ice sheet developed as a single coherent unit (Fig. 18.5).
4. Termination of deformed belts at the margins of one continent and the reappearance of a similar-aged belt, with the same structural trend, on a now-separated continent. Similarly, radiometrically-dated, recognizable continental rock units can be matched across ocean basins (e.g., Fig. 16.21).
5. Magnetic pole positions. Through paleomagnetic studies of the natural remanent magnetizations of iron oxide minerals (e.g., magnetite and hematite) in rocks, it is possible to identify the location of the pole when rocks of a given age were deposited. If continents and magnetic poles had remained fixed, these pole positions, as determined worldwide for rocks of any age, should all plot in the same place, just as observers throughout the world find their compasses now point to only one north pole.

   Although magnetic pole positions are very similar for rocks of the same age in any one continent, positions differ for rocks of different ages on each continent. In addition, the pole position for rocks of a given age is different from one continent to another. These observations are interpreted to mean that continents have moved with respect to the magnetic poles and also have moved relative to each other. By reconstructing the geographic positions of the continents, all the magnetic poles for rocks of a given age can be made coincident (Fig. 18.6).
6. Another magnetic phenomenon, recognized in 1906 by B. Brunhes and called polarity reversal, is used to document sea-floor spreading. At irregular intervals, but averaging about one million years, the earth's magnetic field reverses itself (over a period of up to 10,000 years) so that the present north magnetic pole would be at the other pole (shorter-term reversals, called events, last about 100,000 years). These reversals are well docu-

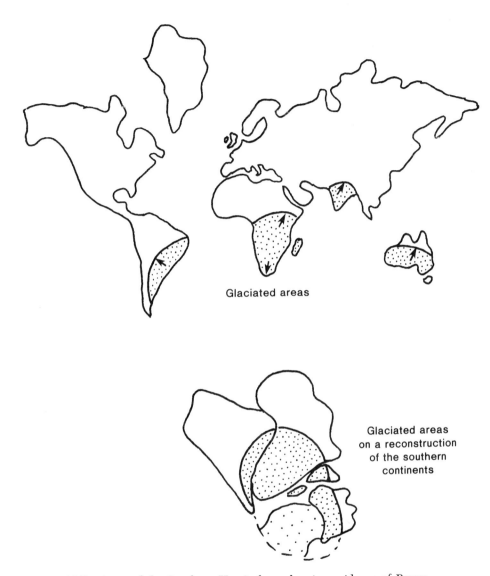

Glaciated areas

Glaciated areas
on a reconstruction
of the southern
continents

Figure 18.5. Areas of the Southern Hemisphere showing evidence of Permo-Carboniferous glaciation (some 275 million years ago). If the continents are reconstructed to form a single landmass (Gondwanaland) the glaciated areas form one ice sheet with flow directions from the center (which does not correspond to the present South Pole).

mented in the oceanic record and are used as a geologic time scale: Flows younger than 0.7 million years show normal polarity, called the *Brunhes normal epoch.* Rocks from 0.7 to 2.5 million years old have reversed polarity and belong the *Matuyama reversed epoch,* which also includes normal polarities at 1 m.y. (Jaramillo normal event) and 2 m.y. (Olduvai normal event). From 2.5 to 3.4 m.y. is the *Gauss normal epoch* (with the Mammoth reversed event at 3 m.y.). Below the Gauss is the *Gilbert reversed epoch.* Almost 200 such magnetic reversals have been recognized in the record of the past 75 million years. As new basaltic lithosphere appears along the spreading oceanic ridges and cools below the Curie temperature, it is magnetized in the direction of the earth's present magnetic field. Lithosphere that was added at an earlier time, when the earth's field was reversed, would have acquired and retained that earlier magnetic direction. The process is repeated continuously. Thus the ocean floor provides a record of the earth's magnetic reversals as alternating strips of differently magnetized lava, symmetrical about the spreading ridges (Fig. 18.3).

7. A related line of evidence is the age of oceanic basaltic rocks. If sea-floor spreading occurs, then rocks farther away from the ridge, on both sides, should be older than rocks nearer the ridge. They are.

8. Similarly, the age of sedimentary layers immediately overlying the basalts is greater away from the ridges (sediments deposited now, rest directly on basalt near the ridge but on successively older layers of sediment away from the ridge).

There are certain features that the plate tectonic theory does not explain fully; for example, deformation occurring *within* plates, poor geographical fit of some areas when landmasses are reconstructed (e.g., Spain). Despite these problems, the plate tectonic model provides an elegant explanation for most geologic phenomena and is truly a "unifying hypothesis".

The driving force of plate tectonics is poorly understood and the present theory is not too different from that advanced for continental drift by Arthur Holmes in 1928: **convection currents** in the mantle (Fig. 18.7).

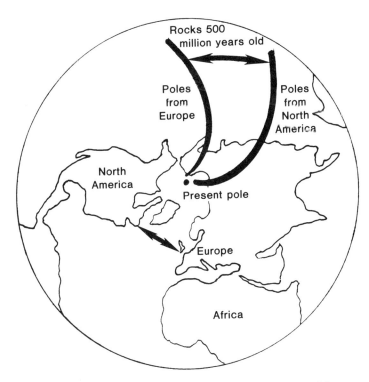

Figure 18.6. The position of magnetic poles as measured for European and North American rocks up to 500 million years old. The two bands can be made coincident if Europe and North America are rotated towards each other. Similar plots can be made for other landmasses.

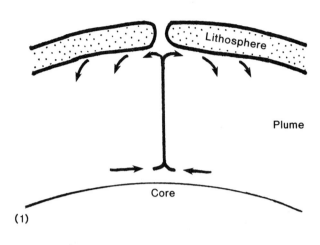

(1)

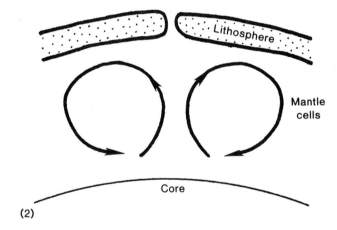

(2)

Figure 18.7. Cross-sections of the mantle showing possible mechanisms for the movement of plates. (1) deep mantle plume, (2) broad convection cells, (3) convection cells restricted to the asthenosphere.

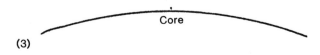

(3)

19. Other Planets

To give perspective on the geologic history of Earth, we summarize the features and evolution of other members of the solar system. The solar system has two types of planet: small, inner (terrestrial) planets like Earth, and giant, outer planets which are gaseous and more like the Sun.

The Moon

The Moon, like Earth, differentiated into a crust, lithosphere, and core and had extensive volcanic activity; it is the same age as Earth, but, except for **meteorite cratering**, has been geologically inactive for the past 3 billion years. In that time no internally generated processes such as volcanism, faulting, or plate collision have operated to produce new topography. The Moon's surface has been unaffected by erosional agents common to Earth (there is no hydrologic cycle because there is no water present). The Moon thus gives us a picture of what Earth might have looked like during the first 1-2 billion years of its history (rocks older than 3.8 billion years have not been identified on Earth, presumably they have been eroded, metamorphosed and/or melted, and incorporated into younger crustal units). Meteorite impact, especially in the early history of the solar system when meteorites were most abundant, is the major process that has modified the Moon's surface. Presumably, meteorites also were significant in modifying Earth's surface appearance (and composition) during its first billion or so years.

The lithosphere of the Moon is relatively simple compared to that of Earth. Materials on the Moon are: a crust of 4 billion-year-old anorthosite (an igneous rock consisting primarily of calcium-rich plagioclase) that makes up the **highlands**; basalt, 3-4 billion years old (generally similar to Earth basalt except it is depleted in low-temperature materials such as water, sulfur, lead) filling the **maria** (dark, low-lying, circular areas). In addition, there is regolith made up of fragments produced by meteorite-impact and material broken down by cosmic rays, etc.

Worth noting also are the small, very deep-focus moonquakes (caused by Earth's gravitational attraction?); **mascons** (or concentrations of mass) in the maria (representing buried meteorites or denser lithosphere feeders to the basalts?); remanent magnetism of the Moon rocks (from a time when the Moon's core was still molten?).

Mars

After the Moon, man probably has been most fascinated by Mars. There is a marked contrast between its northern and southern hemispheres. The **highlands** of the southern hemisphere show extensive meteorite cratering while the **lowlands** of the northern hemisphere have enormous **shield volcanoes** and other volcanic features. Olympus Mons, the largest volcano recorded, is over 500 km (300 miles) wide and is just one of several that stand up to 24 km (15 miles) high (three times the height of Mt. Everest).

Other huge-scale features include the **rifts** and **canyons**. In the equatorial region, one such system is several thousand kilometers long and up to 500 km wide. These features indicate tensional tectonic activity. There is no evidence of compressional features, however, and no indication of plate tectonics.

The surface of Mars has been affected by **surface erosional processes** such as running water, wind (evidenced by dunes), and ice. The permanent polar ice caps are made of water ice. The seasonal expansion and contraction of these polar caps, long noted by astronomers, is caused by carbon dioxide ice. Periodic melting of subsurface ice has given the so-called *chaotic terrain* of ridges and valleys. Large-scale erosion by running water is indicated by channels and tributaries, some of them braided, and other forms analogous to Earth's river systems. The location of the water (other than at the polar ice caps) is a problem. Mars has an atmosphere, largely of carbon dioxide, but the atmospheric pressure is too low to prevent any liquid water from changing to vapor. Perhaps water and CO_2 are stored as ice in the thick regolith and released periodically by volcanic activity or climatic changes.

Mercury

Mercury resembles Mars and the Moon in having cratered highlands and uncratered maria-like basins. Like the Moon, Mercury has a weak remanent magnetic field and has been inactive tectonically for the last 3 billion years. Its high density suggests that its silicate crust is underlain by a large iron core (Mercury's proximity to the Sun should have driven off the easily-volatilized, lighter elements leaving the more refractory ones behind; it has no atmosphere).

Mercury has many scarps (or cliffs), some over 500 km long and 3 km high, that may indicate faulting; these cliffs may be compressional features, the only such features observed outside Earth (Ganymede shows large-scale lateral faulting).

Venus

Venus has surface temperatures close to 500°C and illustrates to an extreme the familiar *greenhouse effect* of carbon dioxide (on Earth, industrial and automobile waste gases are increasing the atmospheric CO_2 content, which, in turn, could cause a warming of Earth's climate). In certain physical characteristics (mass, diameter, density, etc.), Venus is most like Earth but also differs from it profoundly (atmosphere 95% CO_2, high surface temperatures, slow rotation, atmospheric pressures nearly 1 ton per square inch, etc.). Although Venus is cloud-covered, fragmentary information from radar indicates linear mountains, craters, volcanoes, and channels. There are indications of recent geologic activity as photographs show angular fragments of rocks that should have been smoothed by winds and other surface processes if the fragments were old. If geological activity is recent, Venus is unlike the Moon and Mercury.

The Outer Planets

The Outer Planets (Jupiter, Saturn, Uranus and Neptune) are large and massive; they have low densities and very dense atmospheres (including hydrogen, helium, methane). In composition they are more similar to the Sun than to the inner planets. Information from Voyager space-probes indicates that **Jupiter's moons** (satellites) differ from each other, but show evidence of impact cratering (Ganymede, Callisto); linear, fracture-like patterns suggesting tectonic processes (Europa); large-scale lateral faulting (Ganymede) and spectacular, *active* volcanism (Io)—the latter two are the first such phenomena observed outside Earth. Volcanic activity on Io is indicated by numerous caldera-like features and flows, as well as by eruptions of sulfur and sulfur compounds several hundred kilometers high. Ganymede, the largest of Jupiter's satellites, consists of 50% water or ice and 50% rock. Europa has a thin ice crust and a silicate core.

Meteorites

Meteorites fall to Earth from space and are our major source of direct information about at least part of the solar system. They are believed to have originated in the *asteroid belt* (itself the result of collision of small planets beween Mars and Jupiter) or from comets which may have originated near the outer part of the solar system.

The three general types of meteorites are:

1. **Stony meteorites** consisting of silicate minerals (olivine and pyroxene). These meteorites correspond well to what we know about Earth's outer mantle. They have been dated at 4.6 billion years. Two types are recognized: *chrondrites,* with mineral spheroids (chondrules) 1 mm in diameter, and *achondrites* (no chondrules).
2. **Stony-iron meteorites** consisting of silicates and iron-nickel alloy.
3. **Iron meteorites** made up of iron-nickel alloy, that probably are analogous to Earth's core.

Small fragments are burned up in Earth's atmosphere but large meteorites have produced craters over 65 km across (Manicouagan crater, Quebec). Some also suggest that much larger features, such as part of Hudson Bay, are impact craters. Further, it has been proposed that meteorite impacts have produced large lopoliths and have initiated changes in the direction of plate movement, thus affecting orogeny. *Tektites,* small glassy objects first discovered in 1787, may have been produced by meteorite impact on the moon or earth, or by earth's collision with a comet. Their ages generally coincide with earth's magnetic reversals.

From a study of the other members of the solar system we can identify those processes that are common in the system (e.g., impact cratering), rare (active volcanism and lateral faulting), or unique (biosphere, plate tectonics?). We also gain information about the probable early history of Earth; the next step is to consider its physical and biological evolution. A simple outline of that evolution is given in Figure 19.1 as an introduction to *Part II* (see also, Table 4.4).

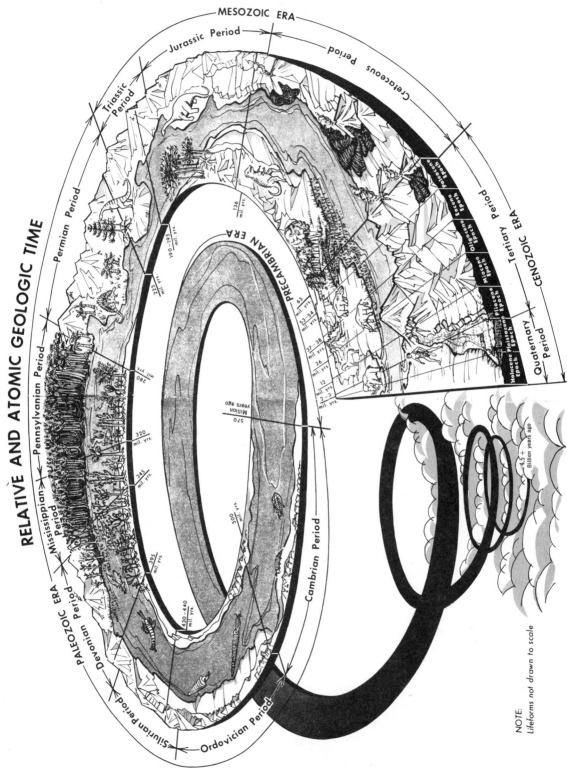

RELATIVE AND ATOMIC GEOLOGIC TIME

MESOZOIC ERA

Jurassic Period

Triassic Period

Cretaceous Period

Permian Period

Paleocene Epoch

Eocene Epoch

Oligocene Epoch

Miocene Epoch

Pliocene Epoch

Tertiary Period

CENOZOIC ERA

Pleistocene Epoch

Holocene Epoch

Quaternary Period

Pennsylvanian Period

Mississippian Period

PALEOZOIC ERA

Devonian Period

Silurian Period

Ordovician Period

Cambrian Period

PRECAMBRIAN ERA

136 mil. yrs.

190-195 mil. yrs.

225 mil. yrs.

280 mil. yrs.

320 mil. yrs.

345 mil. yrs.

395 mil. yrs.

430-440 mil. yrs.

500 mil. yrs.

570 Million years ago

65

53-54 mil. yrs.

37-38 mil. yrs.

26

12 mil. yrs.

2-3 mil. yrs.

4.5+ Billion years ago

NOTE:
Lifeforms not drawn to scale

Figure 19.1. Geologic history expressed in relative and absolute time. (From U. S. Geological Survey.)

Glossary

Names of minerals and rocks are not included in this Glossary; they are given in Tables 2.2, 3.2, 4.2, and 5.1.

For a more complete list of definitions and explanations, see "Glossary of Geology and Related Sciences" published by the American Geological Institute.

Aa flow: Rough-surfaced, blocky lava flow. (Contrast, pahoehoe flow.)

Ablation: Where water loss due to melting, evaporation, etc., exceeds snow accumulation. Characteristic of the lower parts of glaciers (below snowline).

Abrasion: Erosion by mechanical processes; rock fragments moving against other rock material.

Absolute geologic age: Age measured in years, as by radiometric dating methods. (Contrasted with relative geologic age.)

Abyssal: Characteristic of deep ocean basins, with water depths greater than 3700 meters. Features include abyssal plains and abyssal hills.

Accumulation zone: Upper part of a glacier (above snowline) where snowfall exceeds melting. Opposite of ablation.

Aftershock: Earthquake, sometimes several, occurring after a larger earthquake (from the same zone of origin).

A horizon: Topmost layer of soil, including organic material and a leached zone. Overlies the B horizon, the zone of accumulation.

Alluvial fan: Cone-shaped deposit of sediment formed where a river changes slope and velocity. Typically where mountain streams flow onto a broad valley.

Alluvium: Unconsolidated sediment deposited on land by rivers.

Alpha particle: Helium nucleus (2 protons and 2 neutrons) emitted from the nucleus of an element undergoing radioactive decay.

Alpine glacier: See valley glacier.

Andean-type boundary: Collision between oceanic and continental plates producing a volcanic mountain belt and parallel zones of ore deposits.

Andesite line: Line on a map of the Pacific Ocean separating oceanic basaltic igneous rocks from island-arc andesites. Andesites do not occur inside the line.

Angle of repose: Steepest slope angle (measured from the horizontal) at which sediment remains stable.

Angstrom: 10^{-10} meter

Angular unconformity: Type of unconformity where the older and younger rock layers are not parallel. (See also, unconformity, disconformity, nonconformity.)

Anion: Negatively-charged ion.

Antecedent stream: A stream that maintained its drainage pattern through an area uplifted across its path by folding or faulting.

Anticline: Type of fold where the layers form a rough arch with limbs that dip away from each other. When eroded, oldest rocks are in the center. (Contrast, syncline.)

Aphanitic texture: Where the minerals in an igneous rock are too small to be seen without magnification.

Aquifer: Permeable rock or regolith through which enough groundwater moves to supply wells.

Arc-trench gap: Non-volcanic area between an island arc and the adjacent trench, usually at least 100 km wide.

Arête: Knife-like, jagged ridge formed by glacial erosion and mechanical weathering. Typical of areas with valley glaciers.

Arroyo: Steep-sided, flat-floored gully occupied by an intermittent stream.

Artesian well: Well in which water rises under hydrostatic pressure because of a confined aquifer.

Ash: Volcanic material (pyroclastic) less than 4 mm in diameter.

Asthenosphere: World-wide zone in the mantle extending from approximately 100 km to 600 km below the surface. Underlies the lithosphere.

Asymmetrical fold: A fold where the two limbs do not dip into the ground at equal angles.

Atoll: Low coral islands in a ring around a central lagoon.

Atom: Smallest individual particle of an element that has the properties of that element. Consists of protons, neutrons, electrons and other subparticles.

Atomic number: Number of protons in the nucleus of an atom. This number, Z, defines the particular element.

Aulacogen: Linear, graben-like trough produced within a plate during plate separation.

Aureole: Roughly concentric zones of mineral and textural changes produced in rocks metamorphosed by an igneous intrusion.

Avalanche: Slide or fall of a mass of snow. Rock avalanche is a rapid type of landslide.

Axial plane: An imaginary plane that divides a fold into two symmetrical parts.

Backswamp: Marshy part of a floodplain.

Backwash: Flow of water down a beach slope after the uprush (swash) of a breaking wave.

Badlands: Area cut by stream erosion into closely-spaced narrow gullies and ravines; almost no vegetation.

Bar: Ridge of sediment formed offshore or along stream channels.

Barchan dune: Crescent-shaped sand dune with horns pointing downwind. (Contrast, parabolic dune.)

Barrier island: Low, long, narrow, wave-built island, commonly parallel to the shore.

Barrier reef: Reef separated from a landmass by a lagoon.

Base level: The lowest level to which a stream can erode. For most rivers this is sea level; for others, large lakes provide a local base level.

Basin: Tectonic basin is a roughly concentric arrangement of inward-dipping strata, with the youngest in the center. Sedimentary basin is where thick sequences of sediment accumulate. Topographic basin is a depression in the earth's surface.

Batholith: Mass of coarse-grained intrusive igneous rock with an exposed (by erosion) surface area of at least 100 square kilometers. Most are granitic and are confined to continental folded mountain belts. (See also, stock.)

Beach: Deposits of sediment moved by waves.

Bedding: Planar layers (strata), typical of sedimentary rocks, representing changes in depositional conditions. Layers (beds) are separated by bedding planes.

Bed load: Material transported along the bottom of a stream channel.

Bedrock: Continuous solid units of igneous, sedimentary or metamorphic origin. Contrasted with loose, unconsolidated regolith.

Benioff zone: The zone of earthquake foci that extends, from trenches, beneath continents and island arcs. (See also, subduction zone.)

Berm: Nearly-horizontal part of a beach (backshore) separated by the berm crest from the breakers on the beach face.

Beta particle: An electron emitted from the nucleus of an atom undergoing radioactive decay.

B horizon: In a soil profile, the zone of accumulation (lies below the A horizon, the zone of leaching).

Block faulting: Faults that delimit downdropped or uplifted parts of the crust. The faults typically are normal (tensional) type and commonly produce horsts and grabens.

Blueschist facies: Metamorphic mineral assemblage typical of high pressures and relatively low temperatures. Formed during plate collisions. (Compare, greenschist facies.)

Boulder train: A fan-shaped distribution of erratics transported by a glacier.

Bowen's Reaction Series: The order in which minerals crystallize from a cooling magma. By removal of early-formed fractions, it is possible to generate many different rocks from the same parent magma (differentiation).

Braided stream: Stream with many channels that separate and join as the water is unable to transport all the sediment load.

Breaker: Wave that is steepened, becomes unstable and collapses. Characteristic of the surf zone.

Brown clay: Deep-sea accumulation of fine-grained material from rivers and volcanic activity. Also called red clay.

Burial metamorphism: Changes in mineralogy and texture in rocks at the bottom of a thick pile of material. Produced by the temperature and pressure of burial with little or no igneous activity or deformation involved.

Butte: Flat-topped, angular, steep-sided hill formed by erosion of horizontal layers. Smaller than a mesa.

Caldera: Large, steep-walled, roughly circular depression in a volcanic area. Caused by explosion or collapse of the volcanic cone.

Calving: Where large blocks of ice break off a glacier into a body of water (producing icebergs).

Capacity: The amount of sediment a stream can transport past a given point in a given time.

Capillary action: Liquid rising in very narrow openings by surface tension.

Carbonate: A mineral or rock containing the anion group, CO_3. Calcite is the major mineral, limestone the major rock.

Carbon—14: Radioactive isotope of carbon, used in absolute age determinations.

Carbonic acid: A weak acid, H_2CO_3, formed when carbon dioxide (CO_2) dissolves in water. Rain water is a very weak solution of carbonic acid.

Catastrophism: An early geologic doctrine that ascribed geologic changes to sudden, violent, catastrophic events (the opposite of uniformitarianism).

Cation: A positively-charged ion.

Cement: Precipitated material (calcite, quartz, iron oxides, etc.) that holds grains together to form a coherent sedimentary rock.

Chemical sediment: Material formed by chemical and/or biochemical precipitation, usually from sea water.

Chemical weathering: Decomposition of rocks and minerals by the action of water, air, carbonic acid, etc.

C horizon: Lowest part of the soil profile, representing a transition to the parent material.

Cinder cone: Conical, steep-sided accumulation of pyroclastic material around a volcanic vent.

Cirque: Head of a glaciated valley, shaped roughly like an amphitheater.

Clastic: Fragments of pre-existing rocks, minerals and fossils. Clastic texture: fragments held together by a fine-grained matrix and/or chemical cement.

Cleavage: A tendency to break along defined planes. In minerals, the planes are related to bonding in the crystal lattice; in rocks they reflect a preferred orientation of platy minerals.

Col: A gap or low area in a mountain ridge where the headwalls of two cirques erode toward each other.

Columnar jointing: Five- or six-sided columns commonly produced in basalt by contraction on cooling.

Compaction: Reduction in volume and porosity of material, usually due to the pressure of overlying material.

Compensation depth: For calcium carbonate, the ocean depth (approximately 4 km) below which carbonate organisms and sediments are dissolved. For gravity, the depth at which different parts of the crust attain equilibrium in terms of mass and density (isostasy).

Competence: The largest particle a river can transport. (Compare, capacity.)

Composite volcano: A volcano, usually with classic conical shape, that consists of alternating layers of lava and pyroclastics (also called stratovolcano).

Compressional wave: See P wave.

Conchoidal fracture: Smooth, commonly saucer-shaped surface. Typical of obsidian and quartz.

Concordant intrusion: Where an igneous intrusive mass (e.g., sill) remains bounded by the layers of the host or country rock. (Contrast, discordant intrusion.)

Concretion: Localized material precipitated (from groundwater or other pore fluids) around a nucleus within a sedimentary rock.

Cone of depression: Lowering of the water table to form an inverted conical shape around a well. Caused by withdrawing groundwater faster than it can be replaced.

Consequent stream: A stream whose drainage pattern is solely a function of the shape of the land. (Compare, subsequent stream.)

Contact metamorphism: Changes in the mineralogy and texture of rocks around an igneous intrusion (often in an aureole). (Compare, regional metamorphism.)

Continental accretion: The concept that continents have become larger through time by the addition of mobile orogenic belts along their margins.

Continental crust: That part of the crust which contains the continents; typically 30 to 60 km thick, consisting of granitic rocks. (See also, sial.)

Continental drift: Movement of continents relative to each other as the plates on which they rest move as part of plate tectonics.

Continental glacier: Large, thick mass of ice that over-rides the topography (e.g., Greenland, Antarctica). Also called ice sheet.

Continental margins: The area between the continents and ocean basins, consisting of continental shelf, slope, and rise. (Types: active or Pacific or leading, cf. passive or Atlantic or trailing.)

Continental rise: Broad, thick accumulation of sediment between the continental slope and the deep ocean floor.

Continental shelf: Submerged edge of the continent extending, with gentle gradient, to the top of the continental slope (water depth up to about 200 meters).

Continental slope: The steeper-sloped part of a continental margin between the continental shelf and continental rise or trench.

Convection: Transfer of heat through a fluid by rise of hot, less dense material and descent of cold, denser material. Often forms convection cells, or closed paths.

Convergent plate boundary: Collision of lithosphere plates. Site of subduction zone, trench, ore deposits, volcanic and earthquake activity, mountain building. (Contrast, divergent boundary.)

Core: That part of the earth deeper than 2900 km. The outer core is believed to be liquid; the inner, solid iron and nickel.

Coriolis effect: Force tending to deflect north-moving objects to the right in the North Hemisphere (left, for south-moving objects in the Southern) as a result of the rotation of the earth from west to east.

Correlation: Establishing the equivalence of separated exposures of a rock unit.

Covalent bond: Bond between atoms where the outer electrons are shared.

Craton: Part of the continent that has been stable tectonically for a long time. Includes the Precambrian shields of igneous and metamorphic rocks, and the younger sedimentary platforms of the continental interior.

Creep: Slow downslope movement of material under the influence of gravity. Also, slow plastic deformation of rock.

Crevasse: Large, vertical opening in the surface of a glacier.

Cross-bedding: Layers in a sedimentary rock deposited in a non-horizontal position (as in sand dunes or deltas). Also called cross-stratification.

Crust: Outer part of the lithosphere consisting of granitic continental crust (sial) and thinner, basaltic oceanic crust (sima). Separated from the mantle by the Moho.

Crystal: Matter organized into a regular, repeated arrangement of atoms; shows planar faces.

Crystallization: The processes whereby crystals form (e.g., from a cooling magma).

Crystal settling: The sinking of more dense minerals through a less-dense magma, under the influence of gravity.

Crystal system: One of six groups into which minerals can be divided, based on the characteristics of axes drawn through them.

Cuesta: Asymmetrical ridge produced by erosion of gently dipping beds (steep slope on one side, gentle slope on the other). Contrast, hogback.

Curie point: The temperature above which a substance cannot retain magnetization.

Cycle of erosion: The sequence of changes that mark the erosion of an area; the evolution of landscape.

Daughter product: Element produced as the end-product of radioactive decay of another element (the parent).

Debris flow: Rapid downslope movement of regolith, in which water plays a significant role.

Debris slide: Rapid downslope movement of regolith with little or no water involved.

Declination, magnetic: The angle between the rotational and magnetic poles (i.e., between true north and magnetic north).

Decomposition: Term for chemical weathering.

Deep-focus earthquakes: Earthquakes originating at depths greater than 300 km. Typically associated with trenches and plate collisions (subduction zones, Benioff zones).

Deflation: Lowering of ground surface by wind action, involving the removal of clay- and silt-sized particles. A layer of coarse material, deflation armor or desert pavement, is left behind.

Delta: A mass of sediment deposited at the mouth of a river where it enters a lake or the ocean. Water and sediment move over the delta in distributaries.

Dendritic drainage: Map pattern of a river and its tributaries that resembles the branches of a tree.

Density current: Water that flows through the rest of a water mass because it is more dense (turbidity current is a type of density current).

Desert pavement: The result of prolonged deflation whereby only the coarser particles remain (also called deflation armor).

Desiccation: The removal of water from sediment by compaction or evaporation.

Detrital: Sediment transported and deposited by mechanical (physical) processes (in contrast to chemical processes).

Diagenesis: The many processes, physical and chemical, that occur in a sediment during its change to a consolidated rock. (See also, lithification.)

Differential erosion: Different responses to erosion by materials of different resistances; giving steep or gentle slopes, for example.

Differentiation: The various processes whereby a single magma can produce different igneous rocks (crystal settling, for example). Also, planetary differentiation into concentric shells (i.e., core, mantle, crust).

Dike: A discordant, roughly planar, igneous intrusion, typically basaltic in composition. Usually several occur in a dike swarm. (Compare, sill.)

Dilatancy: Increase in rock volume produced by the formation of small cracks during deformation.

Dip: The angle between the unit of interest (e.g., sedimentary layer) and a horizontal plane, as measured perpendicular to the strike.

Discharge: Volume of water per unit of time. Used to measure water flow in a stream.

Disconformity: Type of unconformity where there is evidence of a break in the rock record (usually from fossils) despite the parallelism of the rock layers. (Contrast with angular unconformity.)

Discontinuity: A narrow zone or surface across which the velocity of seismic waves changes rapidly.

Discordant intrusion: An igneous body that cuts through the layers of the surrounding rocks. Dikes are discordant. (Contrast, concordant intrusion.)

Dissolved load: Material carried in solution by a stream.

Distributary: The opposite of tributary—branches receiving water from the main stream, as on a delta.

Divergent plate boundary: A zone where plates separate and new lithosphere material is added, typified by the mid-ocean ridges. (Contrast with convergent boundary.)

Divide: Ridge that separates the drainage basins of two adjacent streams.

Dome: Arching of rock layers, giving a roughly circular map pattern if eroded, with the oldest units in the center. (Contrast with basin.)

Downwarp: A gentle downward movement or subsidence of the crust, without major deformation (see also, epeirogeny). Contrast, upwarp.

Drainage basin: Area, bounded by divides, that is drained by tributaries flowing into a single river.

Drift: All material deposited directly or indirectly by glaciers, including till and outwash. Also used when referring to continental drift.

Dripstone: See stalactite and stalagmite.

Drumlin: Smooth, glacially-rounded hill made of till. Usually many occur together as a drumlin field.

Dune: Mound of sand, shaped by wind or water. Usually applied to wind-blown dunes, which have various shapes.

Dynamic metamorphism: Changes produced in rocks by shearing, grinding, and crushing in fault or shear zones.

Earth flow: Relatively slow movement of waterlogged regolith.

Earthquake: Vibrations in the earth caused by seismic waves produced by sudden release of elastic strain energy along a fault, or by emplacement of magma.

Elastic deformation: Where material regains its original size and shape after deformation.

Elastic rebound theory: Strain energy builds up along a fault until there is slippage along it and the accumulated energy is released as an earthquake.

Emergent coastline: Coastline which is the result of uplift of the land, or lowering of sea level, or both. (Contrast, submergent coastline.)

End moraine: Ridge of till formed at the terminus of a glacier. (See also, terminal and recessional moraines.)

Entrenched meander: Meanders that have downcut because the area was uplifted or base level was lowered (also called incised meander).

Eolian: Wind processes—erosional or depositional.

Epeirogeny: Large, but gradual, vertical movements of the crust without major deformation. (Contrast, orogeny.)

Epicenter: The point on the surface of the earth that is directly above the focus of an earthquake.

Erosion: All the processes that transport rock and regolith.

Erratic: Boulder transported a considerable distance by a glacier.

Esker: Sinuous ridge of glacial material deposited in a stream flowing under the ice.

Estuary: A semi-enclosed area along a coastline where ocean water mixes with river water.

Eugeocline: Similar to eugeosyncline; developed along a continental margin that is not a plate margin.

Eugeosyncline: Oceanward side of a geosyncline, consisting of clastic sediments and volcanics. Distinguished by some from a eugeocline by being involved in plate collision. (Compare, miogeosyncline.)

Eustatic: World-wide change in sea level (as, for example, during advance or retreat of glaciers).

Exfoliation: Weathering phenomenon whereby layers or sheets of rock become detached from the main rock mass.

Extrusive rock: Igneous (volcanic) rock formed when magma (lava) reaches the surface of the earth. (Contrast, intrusive rock.)

Fabric: Orientation of textural components in a rock.

Facies: Sedimentary rocks—the sediment and fossil characteristics that indicate a particular depositional environment (distinguishable from other characteristics). Metamorphic rocks—a particular mineral assemblage; can be taken to indicate the pressure and temperature conditions of metamorphism.

Fault: A break in the crust, usually planar, along which there has been movement of one side relative to the other The surface of rupture is called the fault plane.

Faunal succession, Law of: Evolutionary sequence of changes in plants and animals preserved in the fossil record.

Fiord: A glaciated valley into which the sea has extended.

Firn: Dense compacted snow, also called névé.

Fissure eruption: Extrusion of lava from a linear zone rather than from a central vent (forms ridge, flood, or plateau basalts).

Flood basalt: See fissure eruption.

Floodplain: Bottom part of a river valley consisting of sand and silt (alluvium) deposited during periodic flooding by the river.

Flysch: Sandy and calcareous shales, plus sandstones and conglomerates (graywacke suite or geosynclinal suite).

Focus: Where an earthquake originates (also called hypocenter).

Fold: Bend or warp of a planar feature.

Folded mountain belt: See orogeny.

Foliation: Planar features of metamorphic rocks, including preferred orientation of minerals, and physical breaks called rock cleavage.

Footwall: The block of material below a fault plane. (Contrast, hanging wall.)

Foreset bed: An example of cross-bedding produced at the advancing front of a delta.

Foreshocks: Small earthquakes preceding a larger earthquake.

Fossil fuel: Carbon of biological origin preserved as coal, oil shale, tar sands, oil and gas.

Fractional crystallization: The sequence of formation of crystals from a magma; the removal of early-formed crystals changes the composition of the remaining liquid. (See also, Bowen's reaction series; differentiation.)

Fringing reef: Coral reef attached to a landmass without an intervening lagoon.

Frost wedging: Weathering process caused by the expansion of water when it freezes.

Fumarole: Volcanic vent producing gas, and perhaps water, but no lava.

Gangue: The non-economic part of a mineral deposit. (Contrast, ore.)

Geochronology: The dating of geologic materials and events in both absolute and relative terms.

Geocline: Thick accumulation of sediments on a passive (trailing or Atlantic-type) continental margin. Consists of miogeocline (shelf) and eugeocline (rise). (Compare with geosyncline.)

Geode: A hollow lined with crystals, formed by circulating solutions within the enclosing rock.

Geosyncline: Long, linear trough in which thick piles of sediment accumulate. During plate collision the sediments are deformed, intruded and metamorphosed into a folded mountain range (orogeny). The geosyncline consists of a miogeosyncline near the continent and a eugeosyncline on the ocean side. Differs from a geocline in being on an active (collision or Pacific-type) continental margin.

Geothermal gradient: The increase in temperature with depth in the earth. Averages 2° to 3°C per 100 meters (25°C per kilometer).

Geyser: Groundwater that comes to the surface as hot water and steam. The source of heat probably is deep-seated magma.

Glacier: A mass of ice flowing on a land surface under the influence of gravity.

Glass: Solid matter without the regular internal arrangement of a crystal. Produced when magma is cooled so quickly no crystals can form.

Gondwanaland: Former supercontinent thought to consist of the landmasses now in the southern hemisphere. (See also, Laurasia and Pangaea.)

Graben: A downdropped crustal block bounded by normal faults. (Contrast, horst.)

Gradation: Lowering of the land surface through the action of erosional processes.

Graded bedding: A bed in which coarse-grained particles are at the bottom and fine-grained material is at the top (common in turbidites).

Graded stream: A stream with a smooth longitudinal profile.

Granitization: Formation of granite by metamorphic, essentially solid-state, processes. Some fluid may be present but no large-scale magma is involved. (Granite also results from the crystallization of magma).

Gravel: Coarse sediment consisting of fragments greater than 2 mm diameter.

Greenschist facies: An assemblage of minerals, mostly green, formed by regional metamorphism at relatively low temperatures and pressures. (Compare, blueschist facies.)

Groundmass: The finer-grained material in a rock (usually applied to igneous rocks).

Ground moraine: A blanket of till deposited by a glacier.

Groundwater: Water present in pores or openings in rock or regolith; in the zone of saturation, below the water table.

Guyot: Flat-topped seamount in the ocean basins. Flat top indicates wave erosion when sea level was lower and/or the seamount stood higher (before lithosphere subsided on moving away from the spreading center).

Gyre: The broad circular motion of the surface ocean waters, resulting from winds and the earth's rotation.

Half-life: Time taken for half a given amount of a radioactive isotope to decay to daughter products.

Hanging valley: A tributary that enters the main stream valley at a higher elevation than the stream. Commonly caused by glacial deepening of the main valley.

Hanging wall: The block of material above a fault plane. (Contrast, footwall.)

Hardness: The resistance of a mineral to scratching. A set of minerals with assigned hardness (Mohs' scale) is used to test unknown minerals.

Hawaiian-type eruption: Non-explosive (basaltic) volcanic eruption; forms shield volcanoes.

Headland: A seaward extension of the coastline.

Headward erosion: Erosion whereby the head of a stream advances up and into the regional slope.

Heat flow: The rate at which heat escapes from the earth. High rates are found along spreading plate boundaries.

Hogback: Symmetrical ridge produced by erosion of steeply dipping beds. (Contrast, cuesta.)

Horn: Angular, faceted peak formed by headward erosion of several glaciers around a mountain.

Horst: Relatively high-standing block of the earth's crust typically bounded by normal (tensional) faults. Adjacent blocks are lower (grabens) making the horst a topographic high.

Hot spot: See plume.

Hydration: The addition of water during chemical weathering, whereby different (hydrous) minerals are formed.

Hydrologic cycle: Movement of water in a cyclical path: evaporation from ocean to atmosphere; precipitation to glaciers, rivers, groundwater; flow back to the oceans.

Hydrothermal deposit: Elements, often of economic importance, transported by hot fluids and deposited (as minerals) some distance from their source (which is commonly an igneous body).

Hypocenter: Earthquake focus.

Ice sheet: See continental glacier.

Igneous rock: Rock formed by the cooling of magma.

Incised meander: See entrenched meander.

Inclination: Angle between lines of the earth's magnetic field and a horizontal surface. Changes from essentially 0° at the equator to 90° at the poles.

Intensity: A measure of the effects of an earthquake as indicated by the responses of people and the type of damage caused to structures. Expressed in Roman numerals. (Compare, magnitude.)

Interior drainage: Streams that evaporate and infiltrate before reaching the sea (typical of arid climates).

Intermittent stream: A stream that flows only at certain times (sporadic rainfall in arid climates).

Intermontane basin: A sedimentary basin commonly in a graben, bounded by horsts.

Intrusive rock: Igneous rock formed from magma that crystallizes beneath the earth's surface. Also called plutonic rock. (Contrast, extrusive rock.)

Ion: An atom (or group of atoms) that has an electric charge because it gained or lost electrons. (See also, anion, cation.)

Ionic bonding: Electrostatic bond between oppositely-charged ions.

Island arc: An arcuate belt of volcanic islands associated with plate convergence, oceanic trenches and deep-focus earthquakes.

Isograd: A line on a map joining rocks that have been metamorphosed to the same degree. Based on the first appearance of key minerals.

Isomorphism: When minerals have the same physical appearance but different chemical compositions. (Contrast, polymorphism.)

Isostasy: The concept that parts of the crust move up and down until the masses are balanced. High mountains have deep "roots" of material less dense than adjacent rocks, hence they "float" like an iceberg. (See also, compensation depth.)

Isotope: Variety of an element that has a different number of neutrons (but the same number of protons). All elements have isotopes; some are unstable and change to stable forms by radioactive decay.

Joint: Rock fracture along which there has been no movement. (Contrast, fault.)

Kame: Deposit of stratified drift formed by meltwater as a delta or terrace against a glacier.

Karst topography: Landforms produced by solution of limestone units. Characterized by sinkholes, caverns, underground rivers, etc.

Kerogen: The organic material present in oil shale (and other sedimentary rocks).

Kettle: Depression in glacial material produced by deposition of outwash sediments around a block of ice (which later melts).

Laccolith: Concordant, sill-like igneous body that causes doming of the overlying rocks.

Lagoon: A shallow body of water with restricted circulation to the open ocean.

Lahar: Mudflow involving volcanic ash.

Landform: Feature of the earth's surface shaped by erosional and/or depositional processes. (See also, morphology, topography.)

Landslide: General term for rapid downslope movements of rock and regolith. (See rock slide, debris slide.)

Lateral moraine: Till deposited along the sides of a glacier.

Laterite: Type of soil from which silica is leached, with resulting concentration of alumina and iron oxide. Developed in tropical climates.

Laurasia: Former supercontinent thought to consist of the continents in the present northern hemisphere. (See also, Gondwanaland and Pangaea.)

Lava: Magma that reaches the surface. Forms extrusive igneous rocks; commonly basalt.

Leach: Solution of elements by water moving through the material. (The A soil horizon is the zone of leaching.)

Leachate: Aqueous solution produced by leaching, typically from solid wastes.

Leading edge: The margin of a lithospheric plate (or landmass) that faces in the direction of plate movement (e.g., western South American coast). (Compare, trailing edge.)

Lee slope: See slip face.

Levee: Ridge along the banks of a stream, formed from material deposited during floods.

Lineation: The alignment of any linear features in a rock.

Lithification: Processes, such as compaction and cementation, that change a sediment into a sedimentary rock. (See also, diagenesis.)

Lithosphere: Outermost layer of the earth, 70-150 km thick, consisting of the crust and part of the mantle. Underlain by the asthenosphere.

Load: The total amount of material carried by an erosional agent.

Loess: Non-stratified blanket of silt-sized sediment transported by wind.

Longitudinal dune: Sand dune elongated parallel to the wind direction. (See also, seif dune.)

Longitudinal profile: Elevation of a stream plotted against distance from head to mouth.

Longshore current: An ocean or lake current parallel to the shore, produced by waves approaching the shore at an oblique angle.

Longshore drift: Sediment moved by waves along a beach and in the surf zone.

Lopolith: The largest type of mafic and ultramafic igneous intrusion. Most are saucer-shaped and resemble large sills (except they are discordant).

Low velocity zone: The upper part of the asthenosphere where seismic wave velocities are lower than in the lithosphere above it.

Luster: The way in which light is reflected from the surface of a mineral.

L wave: See surface wave.

Mafic: Magnesium- and iron-rich minerals and rocks. (Basalt is an example.)

Magma: Molten, usually silicate, material that forms igneous rocks when cooled. When extruded at the earth's surface, magma is called lava.

Magnetic reversal: See reversed polarity.

Magnitude: Instrumental measure of the size of an earthquake using the amplitude of seismic waves. Called the Richter scale. (Compare, intensity.)

Mantle: That part of the earth, 2900 km thick, between the core and the crust, consisting of mafic silicates and oxides.

Mare: Dark, roughly circular lunar plain (plural, maria).

Mascons: Concentrations of mass near the lunar surface.

Mass wasting: Downslope movement of rock or regolith under the influence of gravity, commonly aided by water. Also called mass movement.

Matrix: Finer-grained material enclosing coarser material in a rock.

Maturity: Stage in the cycle of erosion where landforms have maximum relief and well-developed drainage patterns, with meandering rivers on narrow floodplains.

Meander: Loop-like development of a river channel, produced by erosion on the outer edge and deposition on the inner part of the bends.

Mechanical weathering: The physical processes that increase the surface area of rocks by breaking them into smaller pieces. Also called disintegration.

Medial moraine: Till deposited in a strip where two lateral moraines join.

Mélange: Jumbled turbidites and ophiolites accumulated in subduction zones (trenches).

Mercalli scale: See intensity.

Mesa: Flat-topped, steep-sided erosional landform of arid climates; larger than a butte.

Metamorphic rock: Rock formed by modifications in mineralogy and texture (in the solid state) due to temperature and pressure changes affecting a pre-existing rock. Recrystallization is a major process.

Meteorites: Metallic or stony material from outer space.

Mid-ocean ridge: Linear, world-encircling, sea-floor tensional feature offset by numerous transform faults. Divergent plate boundary, where sea-floor spreading occurs.

Mineral: Naturally-occurring, inorganic, crystalline compound with definite physical properties and a chemical composition that is fixed within specified limits.

Miogeocline: Shallow, continental side of a geocline. (Compare, eugeocline.)

Miogeosyncline: The shallower, non-volcanic part of a geosyncline. (Compare, eugeosyncline.)

Mobile belts: Linear crustal belts (geosynclines) that have been subjected to mountain building (orogeny).

Mohorovicic discontinuity: Boundary between crust and mantle, identified by a sharp increase in the velocity of seismic waves. Occurs at depths of from 5 to 60 km. Also called *Moho* or M-discontinuity.

Mohs' hardness scale: A series of minerals with assigned hardnesses. From 1 to 10: talc, gypsum, calcite, fluorite, apatite, orthoclase, quartz, topaz, corundum, diamond.

Molasse: Sands, shales and conglomerates. From erosion of an orogenic belt. (Arkoses are common.)

Monadnock: An isolated hill above a peneplain.

Monocline: A type of fold with only one limb, connecting two roughly horizontal sets of strata.

Moraine: Till deposited under, along, or at the terminus of a glacier. (See also, ground, end, terminal, lateral, medial, recessional moraine.)

Morphology: Shape or appearance of the earth's surface. (See also, topography.)

Mountain: An area standing at a higher elevation than the adjacent areas. Larger than a hill. Mountains include volcanoes, fault blocks and orogenic folded belts.

Mountain glacier: See valley glacier.

Mud crack: Crack caused by shrinkage as wet mud dries.

Mudflow: Downslope movement of fine-grained, water-saturated sediment.

Nappe: Part of the crust, commonly showing recumbent folding, moved laterally by thrust faulting.

Neutron: Electrically neutral constituent of the atomic nucleus.

Névé: See firn.

Nodule: Irregular but generally rounded material. Varieties include manganese nodules formed on the ocean floor, and mantle nodules brought up in mafic and ultramafic magmas.

Nonconformity: Type of unconformity where sedimentary layers overlie igneous rocks.

Normal fault: A tensional fault where the hanging wall moves down relative to the other block (footwall). A type of dip-slip fault. (Contrast, reverse fault.)

Nuée ardente: Incandescent, rapidly-moving cloud of volcanic ash and gas. Produced during violent eruptions of Pélean type. Forms welded tuffs.

Obduction: Process during subduction whereby part of the downgoing plate is forced up into and on the other plate.

Oceanic crust: The generally basaltic material, some 5-10 km thick, underlying the oceans. See also, sima; ophiolite. (Contrast, continental crust.)

Oil shale: Fine-grained sedimentary rock rich in organic material. On heating will give liquid and gaseous hydrocarbons.

Old age: Stage in the cycle of erosion characterized by wide floodplains, meandering rivers, a peneplain, etc.

Ooze: Fine-grained, deep-sea (pelagic) sediment containing more than 30% organic material (rest is clay). Separated into siliceous ooze and calcareous ooze depending on the composition of the plant and animal remains.

Ophiolite: Mafic and ultramafic igneous rocks, together with ocean sediment, making up the oceanic crust and part of the upper mantle. Produced at divergent plate boundaries (ridges).

Ore: A mineral deposit of sufficient grade (richness) to be economically feasible. (Contrast, gangue.)

Orogeny: Processes whereby folded mountain ranges are produced, including thrust-faulting, intrusion of batholiths, and metamorphism. Material so deformed is presumed to have been in a geosyncline and involved in plate collision. (Contrast, epeirogeny.)

Outwash: Sediment, usually sorted and stratified, deposited by meltwater flowing from a glacier. (Contrast, till.)

Overturned fold: A fold where both limbs dip in the same direction. As one limb has rotated past vertical, older layers are above younger ones.

Oxbow lake: Crescent-shaped lake. Formed in an abandoned river meander loop.

Pahoehoe flow: A smooth, ropy-surfaced lava flow. (Contrast, aa flow.)

Paleocurrent: Ancient current directions inferred from sediment characteristics (cross-bedding, for example).

Paleogeography: The landforms, position of shorelines, etc., for a given time in the past.

Paleomagnetism: The magnetic properties of ancient rocks and the reconstruction of former pole positions.

Paleontology: The study of ancient life (fossils).

Pangaea: The supercontinent assumed to have existed some 250 million years ago before it broke into the present continents by sea-floor spreading. (It included Laurasia and Gondwanaland.)

Parabolic dune: Sand dune with the general shape of a parabola (also called U-shaped dune). The cusps point into the wind direction. (Contrast with barchan dune.)

Partial melting: Derivation of a magma by melting only part of the available solid material. Minerals with the lowest melting temperature will form liquid first, thus the extent or degree of partial melting controls of the composition of the magma.

Pedalfer: Soil typical of moist climates, having a B horizon rich in clay and iron oxides.

Pediment: A flat, sloping bedrock surface produced by wind and water erosion. Typical of arid, mountain regions.

Pedocal: Soil typical of fairly arid climates. A and B horizons are rich in calcium.

Pegmatite: An igneous rock consisting of very large crystals. Typically composed of quartz, feldspar and mica; intruded as irregular, sub-planar sheets.

Pelagic sediment: Material that accumulates from the water column in the deep ocean basins (clays and oozes).

Peléan-type eruption: The violent type of volcanic eruption that produces a nuée ardente.

Peneplain: The end of the cycle of erosion where the land has a smooth, broadly-rolling topography (characteristic of the old-age stage).

Perched water table: Body of groundwater separated from the main water table by a local, impermeable layer.

Permafrost: Permanently frozen ground.

Permeability: Capacity of material to transmit fluids through pores or fractures.

Petrification: Process whereby organic material is replaced by minerals and converted to rock (by groundwater circulation).

Petrology: The study of rocks; their origin, composition, texture, structure and history.

Phaneritic: Where individual crystals in a rock are visible without magnification.

Phenocryst: Crystal (in an igneous rock) larger than the groundmass. (See, porphyritic texture.)

Piedmont glacier: Where two or more valley glaciers coalesce.

Pillow lava: Form developed when lava flows under water, resembling a pile of pillows.

Placer: An accumulation of heavy, durable minerals having economic importance, found as part of sand and gravel deposits produced by rivers, waves, etc.

Plastic deformation: Where material changes its shape without rupturing.

Plateau: A broad, flat area having a higher elevation than adjacent regions.

Plateau basalt: See fissure basalt.

Plate tectonics: The behavior and characteristics of the earth's lithosphere as explained by the movement and interactions of its twelve or so pieces or plates.

Playa lake: Intermittent lake occupying an arid basin (playa).

Plinian-type eruption: Extremely violent volcanic eruption (Vesuvius, 79 A.D.).

Plucking: Removal of bedrock fragments by the lifting action of glacial ice.

Plume: Confined, pipe-like movement of partially-molten material from the deep mantle up to the lithosphere. Proposed to explain igneous activity within plates. (Also called hot spot.)

Plunging fold: A fold with its axis inclined to the horizontal.

Plutonic rock: Igneous rock formed below the earth's surface; also called intrusive igneous rock. (Contrast, extrusive igneous rock.)

Pluvial lake: A lake that existed when the climate was wetter than the present, common during glacial periods.

Point bar: Arcuate accumulation of sand and gravel deposited on the inside (low velocity part) of a meander loop.

Polarity epoch: A long period during which the earth's magnetic field remained with the same polarity. Shorter periods are called *polarity events.*

Polar wandering: Originally used in the sense that the outer part of the earth had moved with respect to the rotational poles (or vice-versa). Now used more in relation to the apparent movement of magnetic poles. These apparent movements observed in ancient rocks are thought to be the result of movements of the crust itself by plate tectonics.

Polymorphism: The appearance of a given chemical composition in two or more different physical forms (e.g., carbon as either diamond or graphite). (Contrast, isomorphism.)

Porosity: Percentage pore space (voids) in rock or regolith.

Porphyritic texture: Igneous rock texture where some crystals (phenocrysts) are larger than the rest. Indicative of at least two different crystallization (cooling) episodes in the magma.

Primary coast: Coast where the shape is due mainly to non-marine processes (land derived sediments, volcanic activity, tectonism). (Compare, secondary coast.)

Proton: Positively-charged particle in the atomic nucleus. The number of protons (atomic number) determines the element.

P wave: Primary or compressional seismic wave. Fastest seismic wave. (Compare, S wave.)

Pyroclastic: Fragmental material produced by volcanic activity. Includes ash, blocks, bombs, etc. Also called tephra.

Radioactivity: The decay of certain atomic nuclei into other nuclei by the release of alpha and beta particles and energy.

Reaction series: See Bowen's reaction series.

Recessional moraine: Mound of till produced at the terminus of a glacier as it pauses during a general retreat.

Recharge: The addition of water to an aquifer.

Recrystallization: Growth of minerals in the solid state, as a result of heat, pressure, and fluids acting on pre-existing minerals. Common during metamorphism.

Rectangular drainage: Streams showing abrupt changes in direction. Caused by joints or faults.

Recumbent fold: An extremely overturned fold where both limbs are nearly horizontal. Commonly associated with thrust faults and nappe structures.

Redbeds: Sedimentary rocks, usually sandstones, stained red by iron oxide. Generally indicative of subaerial deposition.

Reef: A marine structure built of organic remains, especially coral.

Refraction: See wave refraction.

Regional metamorphism: Changes in mineralogy and texture produced in rocks over a wide area; the result of elevated temperatures and pressures. Accompanied in many cases by igneous intrusions and deformation. (Compare, contact metamorphism.)

Regolith: Unconsolidated material, typically derived from and overlying bedrock. That part of regolith which can support plants is called soil.

Rejuvenated stream: The development of more youthful features in a river system. Commonly the result of uplift of the land or lowering of sea level (base level) causing more active downcutting of the stream.

Relative geologic age: Age of geologic materials, events or features expressed with reference to a standard scale whereby they can be shown to be older or younger than others, without the use of years. (Contrast, absolute age.)

Relief: Maximum difference in elevation within an area.

Remanent magnetism: Permanent magnetism in a rock. Includes thermoremanent magnetization of igneous rocks as they cool; and depositional and chemical remanent magnetization of sedimentary rocks.

Replacement: Where one mineral takes the place of another, either by solid-state ionic diffusion or by one mineral dissolving to provide a depositional site for the other.

Retrograde metamorphism: Rock changes produced during falling temperatures and pressures.

Reversed polarity: Earth's magnetic field when the polarity was opposite to that of the present.

Reverse fault: A compressional fault where the hanging wall moves up relative to the other block (footwall). A type of dip-slip fault. (Contrast, normal fault.)

Richter scale: See magnitude.

Ridge: See mid-ocean ridge.

Rift valley: Valley formed by tension, typically a large graben-like trough.

Rip current: The localized, rapid movement of water from the shore back out to sea.

Ripple marks: Small wave-forms produced in sediment and commonly preserved in rock. The result of river flow, wind, beach or ocean processes.

Roche moutonnée: A rounded knob of bedrock with a gentle, striated side and a steeper, angular, plucked side, produced by glacial advance up the gentle slope.

Rock: An aggregate of minerals that is solid and coherent. The three main types are igneous, sedimentary, and metamorphic.

Rock cycle: That part of the total geologic cycle which emphasizes the rocks developed: Igneous rocks are weathered to regolith, which becomes sedimentary rock, changes to metamorphic rock, eventually melts to produce magma and hence new igneous rocks, etc.

Rockfall: The most rapid downslope mass movement; essentially a free-fall of rock fragments.

Rock flour: Very fine-grained material produced by glacial abrasion.

Rock glacier: Debris held together by ice, flowing under gravity.

Rockslide: Units of bedrock moving rapidly downslope. (See also, debris slide.)

Runoff: That part of precipitation which flows over the ground surface. (Contrast with groundwater).

Saltation: Movement of sediment by small jump-like or bouncing motions.

Salt dome: A structure wherein rock layers are tilted upwards by the intrusion of salt which flows plastically under pressure (from deep layers of rock salt). Of significance as an oil and gas trap and a source of sulfur. Also called salt diapir.

Sand: Fragments ranging from 1/16 to 2 millimeters in diameter.

Scarp: A cliff.

Schistosity: The parallel arrangement of platy minerals in a metamorphic rock. A type of foliation coarser than slaty cleavage.

Seafloor spreading: The creation of new lithosphere at divergent plate boundaries, thus separating the plates. (See also, mid-ocean ridge.)

Seamount: Isolated volcanic mountain on the ocean floor. Some occur in linear belts. (See also, guyot.)

Sea stack: Part of a cliff isolated from the rest by wave erosion.

Secondary coast: Coast where the shape is due mainly to marine processes. (Contrast, primary coast.)

Sedimentary rock: Rock formed by surface processes from fragments of pre-existing rocks, chemical precipitates, and/or accumulated organic material.

Seif dune: A type of large longitudinal sand dune that has been modified by cross-winds.

Seismic: Pertaining to earthquakes and the waves (P, S, and L) produced by them.

Seismic discontinuity: See discontinuity.

Seismograph: An instrument that records the passage of earthquake waves. The record is shown on a seismogram.

Sheeting: See exfoliation.

Shield: A large stable part of a continent, consisting of very old, mainly igneous and metamorphic rocks. With the stable platform, it constitutes the craton.

Shield volcano: Broad, gently-sloping volcanic cone typical of quiet, basaltic, Hawaiian-type eruptions.

Shore: Area between low and high tide. In lakes, the strip of land adjacent to the water.

Sial: A term, contracted from silicon and aluminum, for the granitic, continental-type crust. (Contrast, sima.)

Silicate: The largest group of minerals, consisting of silicon and oxygen tetrahedra bonded to other tetrahedra and other elements.

Sill: A concordant, commonly fairly horizontal, igneous intrusion. (Contrast, dike.)

Sima: A term, contracted from silicon and magnesium, for the basaltic, oceanic-type crust. (Contrast, sial.)

Sinkhole: Feature of Karst topography where the roof of a limestone cavern collapses, creating a surface depression.

Slaty cleavage: A type of foliation where very small platy minerals in a metamorphic rock are arranged in parallel sheets. The rock will split along these sheets, which usually are developed perpendicular to compression and at an angle to bedding planes.

Slickensides: Scratched (striated) and polished surfaces produced by movement along a fault plane.

Slip face: The steeper, down-wind face of a sand dune. Also called lee slope.

Slump: A type of downslope movement where an entire block of material moves as a unit along curved planes of weakness; typically in poorly-consolidated material rather than bedrock.

Snowfield: An area of permanent snow. When it begins to flow, it becomes a glacier.

Snowline: The elevation above which snow remains all year (separating the zone of accumulation from the zone of ablation in a glacier).

Soil profile: A vertical section through a soil showing the horizons (A, B, C) that have developed.

Solid solution: Shown by a mineral when one element takes the place of another so that the mineral's composition varies within certain limits (e.g., plagioclase).

Solifluction: Type of creep, common in cold climates, where part of the regolith is saturated with water while the underlying material remains frozen.

Sorting: The extent to which sediment consists of grains of the same size. (Well-sorted consists of one size; poorly-sorted has a wide range of sizes.)

Spatter cone: Conical, usually small, edifice built from congealed spurts of lava around a volcanic vent.

Specific gravity: The density of material compared to that of water.

Spheroidal weathering: The tendency for fractured and jointed bedrock to weather into rounded boulders because of weathering along and at the intersections of planes.

Spit: A sand ridge, connected to the shore at one end, formed by longshore currents.

Spring: Where groundwater flows to the surface.

Stable platform: That part of the continental interior where old, shield-type material is covered by a veneer of sedimentary rocks (with the exposed shield, it forms the craton).

Stalactite: An icicle-like deposit of calcium carbonate hanging from the roof of a cavern. A type of dripstone.

Stalagmite: A deposit of calcium carbonate, of irregular, conical shape, formed on the floor of a cavern. A type of dripstone.

Stock: An igneous intrusion smaller than a batholith (exposed surface area less than 100 sq. km).

Strain: Change in volume (or a dimension) relative to the original value. Produced by the application of stress.

Strata: Layers of sedimentary rock (singular, stratum). The study of such layers is called stratigraphy.

Stratification: The arrangement of particles or beds in layers.

Stratovolcano: See composite volcano.

Streak: Color of the powder left when a mineral is rubbed on a porcelain plate. Used as a diagnostic physical property of certain minerals.

Stream piracy: The diversion of one stream into another's channel by headward erosion.

Stress: Force per unit area.

Striations: Grooves or scratches produced by glacial abrasion, faulting, etc.

Strike: The compass direction of a horizontal linear feature or a horizontal line in any planar feature. Used, with dip, to define the attitude of strata, etc.

Strike-slip fault: Type of fault where relative movement of the crustal blocks is mainly horizontal. Also called transcurrent or lateral or wrench fault.

Strip mining: A mining method that involves removal of surface material.

Strombolian-type eruption: Explosive volcanic eruption. (Less violent than Vulcanian-type.)

Subduction: Plate collision (convergent plate boundary) where one plate descends under the other. (See also, trench.)

Submarine canyon: Valley cut across the continental shelf or slope.

Submergent shoreline: A coast where sea level has risen, or the land has sunk, or both. (Contrast, emergent shoreline.)

Subsequent stream: A well-adjusted stream that has its tributaries in belts of weak rock, commonly in a trellis pattern. (Compare, consequent stream.)

Subsidence: The lowering of one part of the crust relative to another. Term also used for more local sinking due to withdrawal of groundwater, sub-surface mining etc.

Superposed stream: A stream drainage pattern, developed on rocks of particular type and attitude (now removed by erosion), that is preserved in underlying rocks for which the river pattern is inappropriate.

Superposition, law of: Steno's principle that, in a sequence of sedimentary strata, the oldest layer is at the bottom and the youngest is at the top (unless overturned by deformation).

Surface waves: Relatively slow seismic waves (called L waves) that travel at the earth's surface, not through its interior. (Compare, P and S waves.)

Surf zone: Strip of breaking waves between the outermost breakers and the shoreline.

Suspended load: That part of the total load carried by water or wind, excluding material in solution or rolled along the land surface.

Swash: The uprush of water from a wave breaking on the beach face.

S wave: Seismic wave that travels slower than P wave and is not transmitted through liquids. (Called secondary, shear or transverse wave.)

Symmetrical fold: A fold where the two limbs dip at the same angle (in opposite directions).

Syncline: A type of fold where the limbs dip toward each other. The youngest rocks are in the center of the eroded fold. (Contrast, anticline.)

Talus: Angular fragments produced by mechanical weathering, typically in piles at the base of a cliff.

Tectonics: The study of the deformation and structure of the lithosphere.

Tephra: See pyroclastic.

Terminal moraine: Mound of till deposited at the position of a glacier's greatest advance.

Terrace: A roughly horizontal surface bordered by a steeper slope.

Terrestrial planet: Planets most similar to Earth: Mars, Venus, Mercury.

Terrigenous: Derived from land; used to distinguish ocean sediment of continental origin from material that originated in the ocean basins.

Texture: Size, shape and mutual arrangement of the constituents in a rock.

Thermoremanent magnetism: Magnetization of igneous minerals as the rock cools below the Curie point.

Thrust fault: A type of reverse fault where the fault plane is less than 45° from horizontal.

Till: Unsorted, unstratified material deposited directly by glaciers. (Compare, drift.)

Tombolo: A beach that connects the coast to an island.

Topography: The shapes and elevations of the earth's landforms. (See also, morphology.)

Trailing edge: The margin of a lithospheric plate (or landmass) that faces away from the direction of plate movement (e.g., eastern U. S. coast). Compare, leading edge.

Transform fault: A type of strike-slip that allows one plate to slide past another; typically offsets divergent plate boundaries (ridges).

Transverse dune: Sand dune oriented with its long dimension perpendicular to the wind direction.

Transverse earthquake waves: See S waves.

Trap: A structural or stratigraphic arrangement of permeable and impermeable layers such that oil and gas are localized.

Trellis pattern. A stream drainage pattern where roughly parallel tributaries are developed in layers of softer rock (separated by more resistant layers). See also, subsequent stream.

Trench: Deepest part of the ocean basins. Long, narrow depression marking a convergent plate boundary. (See also, Benioff zone, subduction zone.)

Tributary: A stream flowing into another larger one. (Contrast, distributary.)

Tsunami: Long-wavelength, seismic sea waves that can travel up to 800 km/hr (erroneously called tidal wave).

Turbidity current: A mixture of sediment and water, denser than the adjacent water mass, that thus can flow downslope into deep ocean basins. (Turbidites are graded beds of sand, silt and clay produced by turbidity flow.)

Ultramafic: Igneous rock with low silica and high magnesium and iron content (peridotite is an example).

Unconformity: A surface, separating two parts of a rock sequence, which indicates that an interval of time is not represented by rocks but by an erosional event. (See also, angular unconformity, disconformity, nonconformity.)

Uniformitarianism: The concept that the processes acting today are the same as those that have acted throughout the earth's history, although the rates and intensities may have varied. Thus, "the present is the key to the past."

Upwarp: Area uplifted in a broad arch, without faulting. (See also, epeirogeny.) Contrast, downwarp.

U-shaped dune: See parabolic dune.

Valley glacier: A glacier flowing in a mountain valley. Also called Alpine or mountain glacier.

Varve: Thin sedimentary layer, coarse and light at the bottom, fine and dark at the top, representing one year's deposition in a lake that is frozen in winter.

Ventifact: A rock faceted by the sandblasting action of wind (gives evidence of ancient wind patterns).

Vesicles: Holes formed by gas bubbles in a volcanic rock.

Viscosity: Resistance to flow (usually applied to fluids). Higher viscosity means greater resistance.

Volatiles: Gases or material that can be vaporized easily.

Volcanic bomb: Streamlined fragment of lava (larger than 6 cm) ejected from a volcano.

Volcanism: The processes whereby materials (magma, fragments, gas, etc.) are transferred from the earth's interior to the surface.

Vulcanian-type eruption Very violent volcanic eruption (more violent than Strombolian-type).

Wadi: See Arroyo.

Water gap: Where a stream flows through a more resistant unit, causing the river valley to become narrower.

Water table: The upper surface of the groundwater zone of saturation.

Wave base: The lower limit of wave erosion and transportation in a water body; water depth less than half the wavelength.

Wave-built terrace: Wave-deposited accumulation of sediment (commonly lying on an erosional surface called the wave-cut terrace).

Wave-cut terrace: Erosional surface cut by waves (also called wave-cut platform or bench). May be overlain by wave-built terrace.

Wave height: Vertical distance between wave crest and trough (twice the amplitude).

Wavelength: Horizontal distance between successive wave crests (or troughs).

Wave period: Time for successive wave crests to pass a fixed point.

Wave refraction (water): The turning or bending of a wave as part of it slows on entering shallow water.

Weathering: Chemical and physical breakdown of rocks and minerals at atmospheric temperatures and pressures, with little or no transportation of material.

Wilson cycle: The opening and later closing of an ocean basin. Named for the Canadian geologist, J. Tuzo Wilson, who first proposed it.

Wind gap: A gap through a ridge where a stream used to flow.

Yazoo stream: A tributary flowing alongside the main stream on a floodplain.

Youth: A stage in the cycle of erosion where dissection of the land has just begun, thus narrow V-shaped valleys are separated by broad flat divides.

Zone of aeration: Zone above the water table where the rock or regolith pores are filled with air (and minor amounts of water moving down to the zone of saturation).

Zone of saturation: Zone below the water table where all rock or regolith pores are filled with water.

Index

*Indicates definition is in the Glossary

241

*Indicates definition is in the Glossary

*Indicates definition is in the Glossary

*Indicates definition is in the Glossary